ROUTLEDGE LIBRARY EDITIONS:
ACCOUNTING

Volume 14

DEPRECIATION AND CAPITAL
MAINTENANCE

DEPRECIATION AND CAPITAL MAINTENANCE

Edited by
RICHARD P. BRIEF

Routledge
Taylor & Francis Group

LONDON AND NEW YORK

First published 1984 by Routledge

2 Park Square, Milton Park, Abingdon, Oxon OX14 4RN
711 Third Avenue, New York, NY 10017, USA

Routledge is an imprint of the Taylor & Francis Group, an informa business

First issued in paperback 2016

British Library Cataloguing in Publication Data
A catalogue record for this book is available from the British Library

ISBN: 978-0-415-53081-1 (Set)
ISBN: 978-0-415-70783-1 (Volume 14) (hbk)
ISBN: 978-1-138-96741-0 (Volume 14) (pbk)

Publisher's Note
The publisher has gone to great lengths to ensure the quality of this reprint but points out
that some imperfections in the original copies may be apparent.

Disclaimer
The publisher has made every effort to trace copyright holders and would welcome
correspondence from those they have been unable to trace.

Depreciation and Capital Maintenance

Edited by Richard P. Brief

GARLAND PUBLISHING, INC.
NEW YORK & LONDON   1984

ACKNOWLEDGMENTS

The articles in this volume are reprinted by permission of the authors and the following: Graduate School of Business, University of Chicago: "A Late Nineteenth Century Contribution to the Theory of Depreciation." American Statistical Association: "A General Mathematical Theory of Depreciation." *Accountancy, The Journal of the Institute of Chartered Accountants in England and Wales*: "Depreciation, Cost Allocation and Investment Decisions." University of Chicago Press: "Annual Survey of Economic Theory: The Theory of Depreciation." North-Holland Publishing Company: "The Depreciation Multiplier." The Institute of Management Sciences: "On the Convergence of Periodic Reinvestments." Copyright © 1967 by TIMS. *The Accounting Historians Journal*: "Hicks on Accounting." International Statistical Institute: "The Measurement of Capital." The International Institute of Public Finance, Wayne State University Press: "The Concept of Income in Relation to Taxation and to Business Management."

Library of Congress Cataloging in Publication Data
Main entry under title:

Depreciation and capital maintenance.

(Accounting history and the development of a profession)
1. Depreciation—Addresses, essays, lectures.
2. Plant maintenance—Accounting—Addresses, essays, lectures. 3. Industrial equipment—Maintenance and repair—Accounting—Addresses, essays, lectures.
I. Brief, Richard P., 1933–   . II. Series.
HF5681.D5D45   1984   657'.73   83-49437
ISBN 0-8240-6304-X (alk. paper)

For a complete list of the titles in this series
see the final pages of this volume.

The volumes in this series are printed on
acid-free, 250-year-life paper.

Printed in the United States of America

Contents

Introduction

Of the nine articles reprinted in this short volume, those by Ladelle, Hotelling and Anton are recognized as being the classic articles on the depreciation of a single "machine." Each of these articles was published in a journal that often is not accessible and they are being reprinted here to bring them together in one place.

All three articles view depreciation in present-value terms. In 1916, Paton and Stevenson[1] called this approach to the depreciation problem the "capitalization of future revenue method." Since then, present-value depreciation has been described in many ways although it is most frequently called economic or Hicksian depreciation.

Preinreich's article on depreciation was called an "important contribution" to the subject by Edwards and it is also a classic.[2] However, this article, and his related work, has not received the attention it deserves.[3]

Preinreich criticizes the single machine approach to depreciation and argues that the problem must be formulated in terms of the "continuous flow of productive assets through a plant or production center." In this framework, Preinreich studies growth and changing replacement costs in terms of their effects on various book value, depreciation and profit relationships. Edwards points out that this work seems to have been overlooked by those economists who studied some of these questions in the 1950s. These economists were interested in the relationship between depreciation and replacement in a growing economy.[4]

The next two articles in this book deal with a related problem and it concerns the following question: What happens to the stock of fixed capital if in each period an amount equal to the depreciation in a period is reinvested into identical, perfectly divisible machines? The answer is that the level of capital stock will converge to a limiting value which is a multiple of the initial investment. The multiple is called the "depreciation multiplier." The explanation for the multiplier effect of depreciation is that the provision for depreciation in a period will be excessive in the sense that it will exceed, either in physical or value terms, the amount or portion of the investment that has actually worn out until the limit is reached. Although the idea of the multiplier effect of depreciation has a long history[5], most discussions of depreciation and the related problem of capital maintenance ignore the depreciation multiplier by assuming that the cumulative provision for depreciation will provide "funds" which will, by the end of the machine's life, be equal to the original cost of the machine.

The two articles on the depreciation multiplier in this volume deal with analytical aspects of the subject. Ijiri's paper derives a simple expression for the depreciation multiplier in terms of any depreciation method, assuming that the machine is of the one-hoss shay variety. A one-hoss shay remains completely intact until it goes to the junk pile. De Wolff, on the other hand, assumes straight-line depreciation is used and he derives the depreciation multiplier under various assumptions about the way a machine deteriorates over time. The depreciation multiplier is an important concept and it should be better known in the literature on depreciation and capital maintenance.

The last section of the book contains two papers by Sir John Hicks and my own essay that surveys the writings of Hicks that have special interest for accountants. The two papers were recently reprinted in collections of Hicks's works, but their inclusion in this volume will be more likely to call these papers to the attention of accountants.[6]

In "The Measurement of Capital," Hicks emphasizes the practical difficulties in measuring capital and states that the measurement of capital is one of the "nastiest jobs that economists have set to statisticians." He argues that correcting accounts for inflation would be a "formidable undertaking" and one could not expect accountants to make price-level adjustments because "if one asks why the company needs a balance-sheet the answer must surely be given in terms of its obligations to its shareholders and to other creditors. It has to show, periodically, what it has done with their money." This emphasis on stewardship led to the conclusion that "the balance-sheet of the company, as explained, is designed to show shareholders what has been done with their money. If it is this which has to be shown, the original cost of the actual assets held is the magnitude that is relevant."

In the 1979 paper, "The Concept of Income in Relation to Taxation and to Business Management," Hicks elaborates on some of his earlier ideas. He once again points out that depreciation is not a "firm figure." But then he asks, "If the true depreciation could be assessed objectively, what relation to the accountant's depreciation would it have? This, we shall find, is by no means a nonsense question."

In both of these papers, Hicks argues that the accountant's measure of depreciation normally overstates the gross investment needed to maintain capital since real capital increases faster than it appears to do on the financial statements since the reinvestment of depreciation allowances is made on continually more favorable terms. This concept of a continuous flow of productive assets, and its implications, is discussed by Preinreich.

Thus, Hicks concludes:

> Yet it seems inevitable that when it is proposed to make corrections to the conventional allowances, to adjust to inflation, the question of whether the conventional allowances were appropriate, even in the absence of inflation, is bound to be raised. So one comes back to the subjective assessment—how much can be *safely* taken out of the business—from which the accountant's procedure had been thought to be an escape. There are many things which inflation throws into the melting pot, and this is one of them.

For many years accountants have dealt with depreciation and capital maintenance as a static problem. It is time to recognize the dynamic aspects. We hope the articles reprinted in this volume will provide a foundation for future work on the subject.

Notes

[1]W. A. Paton and R. A. Stevenson, *Principles of Accounting* (Ann Arbor: The Ann Arbor Press, 1916). Reprinted in 1976 by Arno Press.

[2]Edgar O. Edwards, "Depreciation and the Maintenance of Real Capital," in J. L. Meij, ed., *Depreciation and Replacement Policy* (Chicago: Quadrangle Books, 1961), pp. 46–140.

[3]Preinreich's articles on the subject include: "Valuation and Amortization," *The Accounting Review* (September 1937), pp. 210–226; "The Principles of Public-Utility Depreciation," *The Accounting Review* (June 1938), pp. 149–165; "The Practice of Depreciation," *Econometrica* (July 1939), pp. 235–265; "The Economic Life of Industrial Equipment," *Econometrica* (January 1940), pp. 12–44; "Note on the Theory of Depreciation," *Econometrica* (January 1941), pp. 80–88.

[4]In n. 18 (p. 138), Edwards cites the articles by Domar and Eisner.

[5]Horvat's article gives a short history of the subject. Branko Horvat, "The Depreciation Multiplier and a Generalized Theory of Fixed Capital Costs," *The Manchester School* (1958), pp. 136–159.

[6]"The Measurement of Capital" was reprinted in volume 1 of Hicks's collected essays on economic theory, *Wealth and Welfare* (Cambridge: Harvard University Press, 1981). "The Concept of Income in Relation to Taxation and to Business Management" was reprinted in volume 3, *Classics and Moderns* (Cambridge: Harvard University Press, 1983).

Theory of Depreciation: Single Machine

A Late Nineteenth Century Contribution to the Theory of Depreciation

RICHARD P. BRIEF[*]

The unrecognized dispute in the theory of depreciation is whether too much or too little has been written on the subject. In 1957, Davidson began an article, "Depreciation is probably the most discussed and most disputatious topic in all accounting." [1] In the first (1912) edition of his *Depreciation and Wasting Assets*, P. D. Leake began, "Because it closely concerns Profit and Loss, the subject of Depreciation and Wasting Assets is of universal importance, and yet it has hitherto received little or no systematic attention." [2] One might suppose, then, that during the 45 years between these observations most of the discussion appeared. But this supposition is wrong. In 1890, a remarkable paper by O. G. Ladelle appeared in which the first two sentences were, "Depreciation is a subject on which we have of late heard much. Much has been written, and said, upon the necessity of taking account of it, and also upon the different ways in which this may be done." [3]

To my knowledge, no attention has been paid to this article by Ladelle; I can find no references to it.[4] I contend that Ladelle made a great contribution to the theory of depreciation and that much commentary could

[*] Assistant Professor of Business Statistics, New York University. I am indebted to the Schools of Business, New York University, for a grant to support this research. I also thank the librarians of the American Institute of Certified Public Accountants, New York, New York for their invaluable assistance.

[1] Sidney Davidson, "Depreciation, Income Taxes and Growth," *Accounting Research*, July, 1957, p. 191.

[2] P. D. Leake, *Depreciation and Wasting Assets* (London: Henry Good & Son, 1912).

[3] O. G. Ladelle, "The Calculation of Depreciation," *The Accountant* (November 29, 1890), p. 659.

[4] That Leake was apparently unaware of Ladelle's important work is characteristic of accountants' traditional lack of concern with the professional literature.

27

have been avoided if his article had been given any recognition. The primary purpose of this paper is to accord recognition to Ladelle's work. Two ways to proceed suggest themselves: (1) to quote and paraphrase extensively, intertwining my own comments, or (2) to present Ladelle's paper in its entirety thereby minimizing commentary by me. I choose the latter on the assumption that the 1890 volume of *The Accountant* is not readily available to those who would be interested in examining it. Two subsidiary purposes are (1) to present a few biographical notes about Ladelle, and (2) to attempt to place Ladelle's work in historical perspective.

THE CALCULATION OF DEPRECIATION

By Mr. O. G. Ladelle, A.C.A.

Depreciation is a subject on which we have of late heard much. Much has been written, and said, upon the necessity of taking account of it, and also upon the different ways in which this may be done.

We have heard, that to obtain the true measure of Depreciation of an asset we should take various courses, as *e.g.*:

(1) Take the actual decrease in its market value.

(2) Divide the cost (less residue) by the number of years that it is expected to last, and write off each year the quotient thus obtained.

(3) Write off each year a constant proportion of the balance then standing.

(4) Set aside each year such a constant sum, as, with its interest, will accumulate during the life of the asset to an amount sufficient to replace it.

(5) And yet sundry other ideas are adopted.

but there is, I think, a field still untouched, for these rules are all defective, either from inaccuracy or from being applicable to special cases only; and though theoretical discussions are sometimes apt to be dry, yet I venture to submit, that to us accountants, having to take account of almost every sort and kind of asset, rules for special cases, however practical and useful, are insufficient; and that to be ready in every variety of cases that may come before us, not merely to propose a scheme of our own, but to review the particular system that has been already adopted, to see exactly how far it is true and how far fallacious, and to point out its particular defects and their consequences, it is important to have distinctly formulated in our minds, the abstract theory upon which calculations of depreciation in general should rest, and this in a form so general, as to be universally applicable.

It is therefore, to a discussion of this general theory that I now venture to ask this society to honour me with its attention. It is, of course, seldom convenient to work out this theory in its full detail, but my meaning in saying at the close of my paper that we must revert to the general principle, is, that it is this theory which should form the base of our considerations, and guide our general ideas, in propounding or assenting to, the scheme of depreciation for use in any case that may come before us.

I shall, for simplicity, assume the rate of interest to be constant, and this, I think, will generally suffice; but if it should be desired to deal with varying rates we may do so, by adopting symbols yet more general (a)

(a) for writing i_n for the interest of 1 for n

$$f_n \text{ for } (1 + i_n)$$

$$\text{and } | f_n \text{ for } (f_1 \cdot f_2 \cdot f_3 \cdots f_n)$$

our expression in notes (d) and (g) becomes $b_n - i_n (P - \Sigma v_n) | f_{n-1}$.

Allow me, in opening this discussion, to assume, for convenience of illustration, that a large number of persons, numbered from 1 upwards, purchase as a joint venture, an asset, which they do not propose to use simultaneously, but agree to use for a year apiece, in turn, and that we are asked to determine the method of apportioning the cost between them (b).

(b) for, if we can do this, it is quite clear that the union of all these persons into one firm can make no difference to our calculations, and that the same figures, will give the amounts which that firm ought to charge to each year in its accounts.

Now here it is clear that we must not re-value the asset each year, but must entirely disregard market fluctuations, for each speculator is as much entitled to the benefit of any rise, and liable to the loss of any fall, in the market value of their joint asset, as is any of his fellows; and even though after No. 1 has used it for his year, the market should be so risen that the asset is fairly worth its original cost, yet No. 1 cannot call upon the remaining joint speculators to share that value and that cost amongst them, but must pay his just proportion with the others.

To ascertain what this proportion is, it may be stated that as the purchase price is the present value of the total future enjoyment of the asset, so, it is equivalent to the sum of the present values of its enjoyment during each successive year of its duration—so that if we write

b_n for the value of the enjoyment of the asset during its nth year, and

v_n for the present value of this b_n

then if No. 1 pay v_1, No. 2, v_2 and so on, in each case No. n paying v_n the vendor will receive in the aggregate the exact purchase price, and each speculator will pay exactly his fair share for the use of the asset during his year—and if then we can determine the values of v_n our apportionment is complete. (c)

(c) Let P = Purchase price.
$\therefore P = \Sigma v \infty$ where the symbol Σ is used to denote that we are to sum the series stated, giving to the variable (∞) every integral value from unity to the limit indicated (infinity.)

But it may be that those speculators who are to enjoy the asset in future years, will not care to pay for it now, and they may, therefore, agree that No. 1 shall pay at once the whole cost, and that the others shall recoup him at the end of his year the proportions thus paid on their accounts, together with interest thereon for the interval—these payments to be actually made, however, by No. 2, who will in turn recover from No. 3 the shares of his posterity; and so on until the asset is exhausted, simultaneously with which v_n will vanish.

In this case the cost to each speculator will be, the difference between the amounts paid by and to him, (d) i.e.,

> (d) Writing i for the interest of 1 for one year, and f for $(1 + i)$ so that $f^n =$ the amount of 1 accumulated during n years, No. 1 will pay P but recover $(P - v_1)f$. That is to say, he will have expended v_1 but having also invested throughout his year, $(P - v_1)$ will have earned $(P - v_1)i$ interest on that capital.
>
> Similarly No. 2 will pay $(P - v_1)f$ but recover $(P - v_1 - v_2)f^2$ and so for the general expression No. n will pay $\cdots (P - \Sigma v_{n-1})f^{n-1} \cdots$ but recover $(P - \Sigma v_n)f^n \equiv \{(P - \Sigma v_{n-1} - v_n)f^{n-1}(1 + i)\}$ so that his cost will be $v_n f^{n-1} - i (P - \Sigma v_n)f^{n-1} \cdots$

each year the tenant for the year will expend during his year, the original present value of his year's enjoyment, with interest thereon to date of payment, but will earn from his posterity one year's interest upon the amount that he pays for them (e).

> (e) Viz. upon the original present value of their shares accumulated to the commencement of his year.

Obviously the rates of interest and of discount here employed must be equal, but how are they to be determined?

I submit that since, in such a case, every prudent speculator must, and does, either consciously or otherwise, at the time of making his purchase, carefully estimate the value of the interest upon his money, over the period for which he is about to invest it, and include this estimate as a factor in determining the price he is willing to pay. It is this same estimate which supplies the answer to our present question—i.e., the speculators must at the time of purchase jointly agree upon an estimate of the fair rate of interest to represent the value of their money during the life of the asset, and then base their calculations of discount and interest upon this rate, and adhere to it. It may be that this rate will prove to differ from those actually prevalent in future years, but if so the differences caused, will be, *not* profit or loss *of the years in which they occur*, but an original error of the speculators at the date of purchase, and must be dealt with in their accounts, not as normal depreciation, but as an unexpected variation in the market value of their property.

Similar remarks will apply to fixing the values of b_n. It may be uncertain what the real value of the enjoyment of the asset during future years will be, and it may be very difficult to estimate these values, but undoubtedly, it is upon his estimate of this value and of the interest on his money, that, consciously or otherwise, (*f*)

(*f*) And either for the one year, or for several, in which he may be interested, each prudent speculator fixes the price he is willing to pay, and makes his purchase.

A list, then, of these estimated values of the enjoyment of the asset during each year of its life, is obtainable; and given this, with the actual price paid for the asset, and the rate of interest, the values of b_n can be found by calculation (*g*).

(*g*) For representing this series of estimated values by the compound symbol ab_n where a is an arbitrary constant, and taking y as the life of the asset, we have:*

$$P = \Sigma v_y \equiv \Sigma \frac{b_y}{f^{v-1}}$$

$$b_n \equiv ab_n \times \frac{1}{n}$$

$$aP = \Sigma \frac{ab_y}{f^{v-1}}$$

$$b_n \equiv ab_n P \, \Sigma \frac{f^{v-1}}{ab_y}$$

$$a = \Sigma \frac{ab_y}{Pf^{v-1}}$$

whence, returning to the formula in note (*d*) for

$$v_n f^{n-1} - i(P - \Sigma v_n)f^{n-1}$$

we may now write

$$b_n - i(P - \Sigma v_n)f^{n-1}.$$

and the general expression for the cost to each speculator, becomes a known quantity, payable during his year, and equivalent to the agreed value of his year's enjoyment of the asset, less his year's interest upon the amount paid by him for his posterity.

Here, then, we have the cost which each successive joint speculator should bear, and since it can make no difference whether these speculators be separate traders or be copartners in one firm (*h*)

(*h*) For in any case the amounts chargeable to the respective years will be unchanged.

it is clear that this is also, the true principle upon which we should distribute the cost of, *i.e.*, write off Depreciation from, any asset, of any individual, firm or company (*i*).

---

* Author's note. The equation $b_n \equiv ab_n P \Sigma \frac{f^{v-1}}{ab_y}$ originally appeared as $'' \equiv ab_n P \Sigma \frac{f^{v-1}}{a_y}$. This was probably a typesetting error.

(*i*) In the case of a company we may carry the analogy even closer, by regarding the shareholders as changing from year to year, and the directors as acquiring assets as agents of, and in trust for, the various sets of shareholders whom they will represent in the respective years.

And now, putting this into bookkeeping form, we have the theory for each year working thus:

Bring forward the amount from the previous year (or in first year charge cost).

Deduct the agreed value of the current year's enjoyment (*i.e.*, b_n).

Add a years interest at the agreed rate upon the balance, and carry forward the sum thus shown.

Let us now compare this theory with the rules cited in my preface.

Suppose the asset to be part of the plant of a gasworks, costing £120, and having an estimated life of 20 years, when it will be worth £20 for old material; and that, after 10 years use, the electric light is adopted generally in the neighbourhood, so that the plant is now far too large for the company's use, and the market value of it greatly reduced.

By the first method cited, upon the re-valuation at the end of year 11 there would have to be a great reduction.

Is our 11th speculator to bear the whole of this loss? The idea is clearly preposterous. There is a reduction in the capital value of the joint venture, and the loss must be shared by all.

By the second method, each year would bear £5. But we may well conceive that this plant might be nearly as useful, and productive of revenue in the 10th year as in the first. Yet whilst each of our speculators would pay off £5, No. 1 would in addition, have to keep £115 of capital lying out through his year, and No. 10 £70 only, which is arbitrary and often unfair.

Trying the third method, we should find that approximately No. 1 would pay off £11.3 and invest £109 whilst No. 10 would pay off £4.6 only, and invest £49, thus increasing yet further the arbitrary advantage of No. 10.

If we turn to the fourth method, so often put forth as the true and scientific one, we find no improvement. For though our plant may be *nearly* as useful through the 10th year as the first, it clearly will not be *quite* so—there will be more flaws discovered, and more repairs needed to keep it up to its work. Surely, then, it is not fair that with these increasing expenses to pay, *i.e.*, with less net benefit to each successive speculator, the amounts to be contributed to the cost should be uniform. But if we apply the theory expanded above we shall find none of these objections.

Our estimates may be made on various bases—we may agree to estimate the benefit of each year under normal circumstances only; or we may agree that all chances shall be taken into account, and the values of b_n be reckoned accordingly. (*k*)

(*k*) And similarly any number of special circumstances may be dealt with, each having its appropriate bearing upon the original estimates of b_n.

In the first case we should, in the 11th year write off capital (l)

(l) Or our speculators would all subscribe to a fund, for the use of Nos. 11 to 20.

an amount equivalent to the extraordinary loss that had occurred—not as depreciation, but as a special loss of capital to be dealt with by itself. In the second case Nos. 11 to 20 could not complain. They have agreed to take their chance, and have already had a fair allowance for so doing. In fact, when this second method is adopted we may generally consider the charge born by each year as consisting of two parts, which may be stated either in one sum, or separately, one part being for the enjoyment of the asset under normal circumstances, and the other part, an insurance premium against casualties.

Here Nos. 11 to 20 have received this premium, to cover the risk which has now ripened into loss, and must stand to their bargain. Still not throwing it entirely on to No. 11, but sharing it between them (m) and in practice we may

(m) Either by adhering to their original estimates, or by paying at once to No. 11, such a proportion of the present values of their respective shares, as shall together, amount to the loss sustained. Or in our books—by writing off from the asset as a special item, the amount of the loss; and reducing the charge to each present and future year accordingly.

often prefer to adopt this plan; remembering, that it is safer for future years, to have these premiums in hand, than to have a right to contribution from years which are past and gone. But to proceed, the objections to the second, third, and fourth methods, are entirely met—as to the varying capital employed, by the fair interest allowed upon it; as to their arbitrary character, by our whole system being based upon *reasonable* estimates; and as to the decreasing revenue, by the yearly benefits having been fairly estimated in view of that circumstance.

Having now concluded our discussion of this theory—if it be desired to glance also at the practical application of it in detail we shall see that the main points of it, are, the fixing the rate of interest and the values of the series of b_n.

For the interest—we have simply to settle the rates which may fairly be expected from investments such as we are dealing with, during the life of our asset—being careful, generally, for the sake of safety, to keep the rate low enough. (n)

(n) For since a part of the charge to later years consists of interest earned from them by earlier years; if the rate of this interest be higher than the real value of money, the effect will be, to unduly tax these later years, in favour of the earlier ones. By increasing the rate of interest, the total charge is increased, but the vendor gets the same, and the difference, is profit to the early years, at the expense of the later ones.

Coming however to the values of b_n our task is sometimes much more intricate, for to arrive at a fair estimate of the net value of the future enjoyment of our asset we must consider alike, the chance of ever getting that enjoyment, the income to be derived from it when got, and the outgo to be spent in making that income.

And here we may require the advice of other experts intimately acquainted with the business concerned. (o)

(o) As for example, when our chance depends upon the survival of a life, the life of plant, machinery, or buildings; or the stability of various works, &c.; or where our income and outgo are not constant, but require time for development, have their prime and their decay, are likely to increase or decrease, to be depressed, or augmented, by new inventions; or depend upon the many other special circumstances which affect the revenues from different classes of property—we may have to seek the assistance of actuaries, engineers, architects, merchants, &c.—and frequently those concerned in managing our particular business.

We may frequently too find it desirable to deal with these points separately, and for the purposes alike of convenience in working, and of supplying useful information in our account, to state, as one item, the depreciation based on calculations as to the normal income alone; as another item, an amount based on calculations as to the normal outgo alone; and as a third, a reserve to meet contingencies. (p)

(p) As e.g. with our gas works, we may write off depreciation on the assumption that the plant is kept in constant repair free of expense to us, and then state separately the amount payable to a repairs fund, from which to pay this expense, and those carried to a reserve to provide for casualties.

And when this is done we may conveniently arrange the bookkeeping thus:—

In a ledger ruled with two money columns on each side, open an account from the particular class of asset—say gasometers—a second for depreciation of gasometers, a third for repairs of gasometers, and a fourth for reserve on gasometers. To gasometers' account debit in the inner column the cost of gasometers, extend yearly the total (or balance) of the year's expenditure, and cast the outer column after each extension. To the depreciation account credit each year (by P. & L.) the agreed amount of depreciation, irrespective of repairs.

To the repairs account debit the cost of repairs, extending yearly as before, and credit each year (by P. & L.) the agreed charge for that year's repairs.

To the reserve account credit each year (by P. & L.) the agreed amount of reserve for that year.

In drawing the accounts it may be well to show the balance remaining on gasometers, after adding or deducting the balance of these other accounts, but in the books themselves I prefer to keep them distinct, showing upon the gasometers account the total cost to date, on the depreciation and reserve accounts, the total sums charged or reserved to date, and on repairs account the yearly balance, for I find that this gives a ready view of the position, and is very convenient in working.

If gasometers be realized (or thrown aside) we should credit to gasometers account, the original cost, debit to depreciation account the depreciation that has been charged in respect of them, and then carry to the reserve account the difference between the balance of these two items and the price realized (or value of residue). And in books thus kept this is easily done, for, having each part of the account by itself, we can readily ascertain the cost, and the depreciation charged against it to any given date, and so cease charging at the proper time, whereas when the depreciation is credited yearly to the plant account, and the balance only carried forward, these are not so readily seen, and it is easy to charge depreciation of assets, either in excess of their original cost, or after they have ceased to exist.

The repairs account should balance, or nearly so, at intervals, and if it fail to do so, it may be necessary to assist it by a transfer to the reserve account.

The reserve account will, of course, be a continually increasing credit, subject to such debits as may be occasioned by the casual losses charged against it, but it clearly cannot survive the particular assets to which it is related, so that when they disappear this account too should be carried to a general reserve, or dealt with in other suitable way.

In doing thus, however, we must bear in mind that we are really dealing with different parts only of one subject, and that though we may state them separately, we should consider them together—(q)

(q) In determining our quantities the same arguments and the same formulae will apply. Our symbols may be sometimes positive, and sometimes negative, but we shall be using the same principle, and subject to the same considerations.

and we should be careful, too, to keep our yearly estimates as regular as circumstances will allow. (r)

(r) As e.g. Find some regular progression, applicable to our particular case and then deviate from it as little as possible, for this generally adds greatly not only to our convenience in making our calculations, but also to the readiness with which others can grasp their business import when made, whilst accounts in which such items fluctuate from year to year are generally open to suspicion.

We may frequently find results thus obtained agree with those given by the methods cited in my preface, and particularly with the second of them, and some might thereby be led to think that all these investigations were useless, as we might have adopted at once the simpler method.

But this is fallacious—The results will agree, not invariably, but under certain conditions only. (s)

(s) Where the yearly profit happens, as it so often does, to decrease throughout by constant sums, our results will agree with those of this second method. Where it decreases in a constant ratio, we shall agree with the third method; and where it is stationary, through a given time, and then lapses entirely, we shall agree with the fourth, and sometimes, too, correct results may be obtained from a combination of these rules, for in these particular cases, our formulae happen to be identical with those giving these rules.

We may often see what the conditions are, and apply at once a suitable rule, but except where these particular conditions occur, these particular rules, which are attached to them, are useless, and we must revert to the general principle.

[Editor's Note: This concludes the Ladelle essay. We are indebted to *The Accountant* for permission to reproduce it.]

A Biographical Note

In 1890 Ladelle, 28, had taken first prize in the Final Chartered Accountant Examination in June and had accepted a position with Messrs.

Price, Waterhouse & Co. His paper was to have been presented at a regular meeting of The Chartered Accountants Students' Society of London on October 22, 1890, and it had been circulated among the members for discussion at this meeting; shortly before this date, Ladelle became ill with typhoid fever while on an audit engagement and died on October 31, 1890. According to a headnote in *The Accountant,* "the consideration of the paper was adjourned *sine die.*" Aside from an obituary notice and a comment in the January 3, 1891 issue of *The Accountant* where his unexpected death was noted with regret in a "Retrospect" of 1890, there appears to be no further mention of this paper or its author.

One can only speculate why this noteworthy contribution to the theory of depreciation was overlooked. Perhaps Ladelle's contemporaries did not understand his concepts and he did not live long enough to explain them. Later writers, like Hatfield, who made frequent use of *The Accountant,* probably overlooked this paper which was awkwardly placed in the two issues in which it appeared; since each annual volume contained approximately 1000 pages, this article could easily have been missed.

Historical Perspective

Although Ladelle made no specific references to previous work, a number of articles and several books on the subject of depreciation had been published prior to 1890.[5] These laid the foundation for contemporary depreciation methods, and the stock of knowledge about both depreciation principles and practices has not increased substantially since.

One paper of 1883 contains one of the first expositions of the concept that underlies modern depreciation practice—the going-concern principle.

> Manufacturers and traders do not construct business premises or lay down special plant in the intent of a short period.... A large, irrecoverable outlay is generally expended in preparation of a factory, and unless speedy destruction overtakes any certain business, the method of treatment for annual accounting is one of "going concern." Otherwise, since all machinery may be said to be secondhand when once it has turned round, and perhaps unsaleable then at half its cost..., a depreciation of 50 per cent would probably not be sufficient to represent the realisable value at the end of the first year. Manifestly, the career of the business contemplated must have an assumed term, and the cost of the consumption of machinery and other erections must be attributed to the whole term.[6]

Other concepts can be traced to the period prior to 1890. The practice

---

[5] Numerous articles on depreciation appeared in *The Accountant* during the last quarter of the nineteenth century. Papers written on other topics, e.g., auditing and balance sheets, also were relevant to the depreciation problem. The first book relating to this subject was apparently written in 1870: E. Price Williams, *Maintenance and Renewal of Railway Rolling Stock* (William Clowes & Son, 1870). Fourteen years later the well-known book, Ewing Matheson, *Depreciation of Factories* (London: E. & F. N. Spon, 1884) appeared.

[6] Edwin Guthrie, "Depreciation and Sinking Funds," *The Accountant,* April 21, 1883.

of valuing an asset at original cost less depreciation and the calculation of (systematic and rational) annual provisions for depreciation were intensively discussed by some of the leading accountants of the day. There were, of course, exceptions taken to these views, both in principle and certainly in practice,[7] but depreciation accounting, as it is currently practiced, was soon structured. However, a *theory* of depreciation had not developed; that is, there were no concepts to be used in evaluating specific depreciation practices.

Ladelle saw the question of depreciation as an allocation problem, and he may have been the first to associate the problem of depreciation with the more general problem of allocating joint costs. His theory of depreciation is consistent with the so-called "accounting approach" to depreciation. This approach has been condemned because it abandons the attempt to value assets and accepts many methods of arriving at book values which are considered "systematic and rational." Consequently, it has been argued that this approach is "sterile" because "without a theory of valuation we cannot have a theory of depreciation."[8]

Ladelle, however, argued that depreciation accounting *is* a system of cost allocation not asset valuation. He arrived at this conclusion using scientific rather than pragmatic reasoning, and this is his major contribution to depreciation theory. According to Ladelle, the cost of an asset is joint to the periods during which it is in use, and the allocation of depreciation to each period must be based on the expected net enjoyment to be derived during the period, after adjusting for the *agreed* interest on the unallocated portion of cost. The interest that is earned during the period is equivalent to the "normal" rate of profit.

Gains or losses arising from unexpected changes in market values, interest rates, etc., cannot be allocated to the period in which the unanticipated event occurs. The *capital* gain or loss is joint to all periods and is not assignable to any one period. The gain or loss must be "shared by all."

In effect, Ladelle reasoned that *ex post* asset valuations affect only the capital account. Thus, if all expected flows, both negative and positive, were agreed on at the time of purchase, all differences between *ex ante* and *ex post* flows must be viewed as aberrations, i.e., capital gains or losses, caused by an "original error of the speculators." These *ex post* changes, i.e., unanticipated changes in the value of an asset are capital gains or losses and, in principle, cannot be assigned to the profit of a particular period because they are joint to all periods. Ladelle's theory of depreciation therefore has relevance to profit theory. Some 30 years after Ladelle, Frank Knight explicitly introduced the notion of risk and uncertainty, i.e., error,

---

[7] Richard P. Brief, "Nineteenth Century Accounting Error," *Journal of Accounting Research*, 3 (Spring, 1965); "The Origin and Evolution of Nineteenth-Century Asset-Accounting," *Business History Review*, XL (Spring, 1966).

[8] F. K. Wright, "Towards a General Theory of Depreciation," *Journal of Accounting Research*, 2 (Spring, 1964).

into the economic theory of profit.[9] A number of articles by Knight and others were written on the subject, culminating in a series of papers that appeared in the early 1950's.[10] Many of Ladelle's views can be found in these writings.

---

[9] Frank H. Knight, *Risk, Uncertainty and Profit,* Harper Torchbook ed. (New York: Harper & Row, 1965). First published in 1921.

[10] J. Fred Weston, "A Generalized Uncertainty Theory of Profit," *American Economic Review,* Vol. XL (March, 1950). See also comments on this article and the rejoinder by Weston in *op. cit.,* Vol. XLI (March, 1951). More recently, some of the ideas expressed by Ladelle have been introduced in accounting literature. See, for example, Hector R. Anton, "Depreciation, Cost Allocation and Investment Decisions," *Accounting Research,* April, 1956.

A GENERAL MATHEMATICAL THEORY OF DEPRECIATION

Harold Hotelling

A GENERAL MATHEMATICAL THEORY OF DEPRECIATION[1]

By Harold Hotelling, *Food Research Institute, Stanford University*

In the older treatments of depreciation the cost, or "theoretical selling price" of the product of a machine, was conceived of as determined causally by the addition of a number of items of which depreciation is one. In other words, depreciation was first computed by some rather arbitrary formula not involving the theoretical selling price, which was then found by the addition of depreciation to operating costs and division by quantity of output. It will be shown in this paper that depreciation and theoretical selling price must be computed simultaneously from a pair of equations which are frequently a bit complicated. The differences in the results obtained from the arbitrary and mathematical formulæ are often very large.

The simple methods referred to, which still prevail generally in business, are analogous to the naïve type of economic thought for which the only determiner of price is cost and which fails to consider the equally important rôle played by demand.

The "unit cost theory" gave the first recognition to the reciprocal relation existing between the value of a machine and the value of its product. It also deserves the credit for taking into account operating costs and output which, as in the present paper, are assumed to have been determined from experience. However, this theory in its usual form involves a number of serious errors of reasoning which have been pointed out by Dr. J. S. Taylor.[2]

As an improvement on the unit cost theory Dr. Taylor puts forward a method which, it is fair to say, is the only one that has been proposed which ever gives correct results. To it the present paper owes much. He assumes that unit cost plus (defined as operating cost, plus depreciation, plus interest on the value of the machine, all divided by the number of units of output) is to be determined by the conditions that

(a) it is either to remain constant during the machine's life or is to increase in a specified manner; and

(b) it is to be made a minimum.

This procedure is open to a possible criticism which he does not anticipate, namely, that, since the unit cost plus depends on the distribution of depreciation charges, there is danger of the accountant

[1] Presented before the American Mathematical Society (Chicago meeting), December 26, 1924.
[2] "A Statistical Theory of Depreciation Based on Unit Cost," this Journal, December, 1923, pp. 1010–1023.

deceiving himself into thinking that he has reduced costs when he has merely changed his system of charging depreciation to make the unit cost appear less. If unit cost is to be made a minimum in any useful sense, it is to be done by performing some operation on the machine itself, rather than by adjusting depreciation charges, or by any other process of bookkeeping. The criticism may be answered on Dr. Taylor's behalf by the fact that he uses the test (b) above only to determine the time for the tangible operation of scrapping the machine, and then uses (a) to distribute the depreciation charge over the useful life.

But even if we admit the validity of (a) there remains a question concerning (b). Does the manufacturer desire to make his unit cost plus, in the sense defined, a minimum? Or may not considerations of profit lead him to scrap the machine at some different time from that which makes unit cost plus a minimum, or to shut down his factory or restrict the output of the machine? We shall indeed give a proof that in certain cases the tests (a) and (b) give correct results.

To one important element in the problem I find no allusion in the literature. The value of an old machine or other property must depend upon the operating cost. In all depreciation theories based on unit cost it is assumed that the operating cost is known, and the value is then calculated. But operating cost always includes elements which depend on the value. Hence we cannot know the operating cost until we know the value; that is, until the problem is solved. We shall resolve this difficulty by solving a simple integral equation.

The viewpoint of the present treatment is that the owner wishes to maximize the present value of the output minus the operating costs of the machine or other property. This quantity is, in fact, the value of the property. The first step is, therefore, to set up an expression for the value of the property in terms of value of output, operating costs, and the life of the property. Various hypotheses concerning the economic situation are then applied to evaluate the unknowns appearing in the equation. Depreciation is defined simply as rate of decrease of value. A mathematical formulation of the problem of depreciation is presented from which, it is believed, the whole of the valid portions of existing depreciation theories are readily deducible, and which also lays a foundation for further developments. This treatment is adapted to the consideration of depreciation in connection with variation of such factors as prices and interest rates. The theory is in fact completely general. The property involved will be referred to as a machine purely for convenience of language. The amount of computation involved in a particular application will depend upon the degree of refinement of the data; in the present stage of knowledge of mortality tables of property the work can be made quite simple. A specimen problem is worked out numerically.

Continuous functions, instantaneously compounded interest, and integration will be used as better adapted to the needs of theory than the discontinuous functions and summation processes usually employed. The methods of the integral calculus also appear to me simpler than those of algebraic summation even where only integral numbers of years are involved; while the occurrence of portions of years, which does not seriously affect the integration process, is always a source of trouble when discontinuous functions are used.

Obsolescence is a risk of essentially the same nature as fire, earthquake, or burglary, and should be provided for in the same manner—namely, by an allowance out of operating expenses similar to an insurance premium. Even where no insurance or inadequate insurance is carried, an allowance should be made. Obsolescence is thus to be relegated, with insurance, to the category of operating expenses.

FUNDAMENTAL FORMULA

If a machine with operating cost per year $O(\tau)$ produces $Y(\tau)$ units of output per year at time τ and if the value ("theoretical selling price") of a unit of output is x, the annual rental value of the machine at time τ is

$$R(\tau) = x \cdot Y(\tau) - O(\tau).$$

Now the value of a machine is the sum of the anticipated rentals which it will yield, each multiplied by a *discount factor* to allow for interest, plus the scrap or salvage value, also discounted.[1] In the

---

[1] The "force of interest" $\delta(t)$ is defined as the rate of increase of an invested sum s divided by s

$$\delta(t) = \frac{1}{s}\frac{ds}{dt}.$$

It follows by integration that $s = s_0 e^{\int_0^t \delta(\nu)d\nu}$, where s_0 is the value of the invested amount when $t=0$ and ν is a mere variable of integration. Hence $s_0 = s e^{-\int_0^t \delta(\nu)d\nu}$. That is, the present value s_0 of a payment to be made t years hence is the amount s of the expected payment multiplied by the discount factor $e^{-\int_0^t \delta(\nu)d\nu}$. If $\delta(\nu) = \delta$, a constant, the discount factor is $e^{-t\delta}$. The discount factor takes the form v^t if we write $v = e^{-\delta}$; and $v = 1/(1+i)$ in this case, where i is the *rate* of interest in the ordinary sense, *i.e.*, the interest payable at the end of a year on each dollar invested for the year. Then

$$\delta = \log e(1+i) = i - \frac{i^2}{2} + \frac{i^3}{3} - \ldots$$

and

$$i = \delta - 1 = \delta + \frac{\delta^2}{2!} + \frac{\delta^3}{3!} + \ldots$$

Since both these series converge rapidly they afford a ready means of passing from force of interest to rate of interest or vice versa. The difference between i and δ is ordinarily very small.

The value at time t of a payment to be made at time τ is evidently the amount of the payment multiplied by $e^{-\int_t^\tau \delta(\nu)d\nu}$. If s is constant, this discount factor equals $v^{\tau - t}$.

most general case the rate of interest will vary with the time. If $S(n)$ is a function giving the scrap or salvage value at the time n at which the machine is to be discarded, the value at time t is given by the following fundamental formula:

$$V(t) = \int_t^n [xY(\tau) - O(\tau)]e^{-\int_t^\tau \delta(\nu)d\nu} d\tau + S(n)e^{-\int_t^n \delta(\nu)d\nu}, \qquad (1)$$

τ and ν being variables of integration representing time. In case the interest rate is constant (1) becomes

$$V(t) = \int_t^n [xY(\tau) - O(\tau)]v^{\tau-t} d\tau + S(n)v^{n-t} \qquad (1a)$$

Since the value of a new machine is its cost c, we have for $t=0$,

$$c = V(0) = \int_0^n [xY(\tau) - O(\tau)]e^{-\int_0^\tau \delta(\nu)d\nu} d\tau + S(n)e^{-\int_0^n \delta(\nu)d\nu}. \qquad (2)$$

Having once bought the machine, the owner wishes to conserve its value as far as he can. This, in the light of (1), is the same as saying that he hopes to get as much net rental as possible out of the machine, interest and scrap value being considered. We, therefore, adopt as the basis of our further work the simple postulate:

    I. *Everything in the owner's power will be done to make $V(t)$ a maximum when $t > 0$.*

A second postulate of less general validity, which is often assumed tacitly or is thought to follow from I, is the following:

    II. *The machine is always operated at full capacity.*

We shall assume II tentatively.

We suppose that c and the functions Y, O, S and δ of the time are known; but before we can evaluate the right member of (1) we must make some hypothesis which will enable us to determine n and x. There are several possibilities.

(A) If the property has a known useful life n, (2) is a simple equation in x alone.

(B) If x is known, the most obvious way to find a value for n is to solve (2). This, however, can be justified only in very special circumstances. For if x is known in advance of the solution of our problem of depreciation, it must be determined by competitive conditions entirely beyond the control of the machine's owner. He may, for example, use the machine to do work usually done by some other process. In such a case (2) does not hold unless ideally fluid conditions of competition have acted to bring the price c of the machine to such a point as to justify its purchase when n is determined to best advantage. Thus to use (2) to find the best life to allow the machine, the owner must rest precariously upon the judgment of his competitors.

The correct procedure is as follows. Let m be the time, if it exists (if not, put ∞ for m), at which the output of the machine ceases to be worth the operating expenses. Then m is determined by the equation $xY(m) - O(m) = 0$. We may for the sake of generality consider that x is a given function $x(t)$ of the time. In this case m is determined by $x(m) \cdot Y(m) - O(m) = 0$, which we may also write $R(m) = 0$. The sum of the possible future rentals of the machine, discounted at (say) a constant rate of interest is then $\int_t^m R(\tau)v^{\tau-t}d\tau$. This quantity will decrease until, at time n, it reaches the scrap value $S(n)$. Thus n is determined by $\int_n^m R(\tau)v^{\tau-t}d\tau = S(n)$, where $R(m) = 0$. It is of course possible that these equations may have a plurality of solutions. The correct solution is then the one which makes $V(t)$ greatest.

(C) If, as is most common, both x and n are to be determined, we use the equation (2) connecting these unknowns, solving it simultaneously with an additional equation obtained by means of I.

If all economic conditions are static, x is a constant. If economic conditions are not static but their trend can be estimated, we may consider that the value of a unit of output varies in a specified manner with the time, but that its general level depends upon an unknown parameter α. For example it may be supposed that x increases in geometric progression with known ratio r, but that the initial and final values are unknown. We then write $x = \alpha r^\tau$, or in general, $x = x(\tau, \alpha)$.

Take first the static case. Supposing all our functions continuous and with continuous first derivatives, we differentiate (1) and put $dV(t)/dn = 0$. The result of this operation becomes, after cancelling out the common factor $e^{-\int_t^n \delta(v)dv}$ and solving [1] for x,

---

[1] The derivation of (3) given in the text is strictly accurate only in the case in which x is independent of n. If, as is generally the case, x is a function of n, the result of equating to zero the derivative of the right member of (1) with regard to n is

$$\frac{dx}{dn}\int_t^n Y(\tau)e^{-\int_t^\tau \delta(v)dv}\,d\tau + [xY(n) - O(n) - \delta(n)S(n) + S'(n)]e^{-\int_t^n \delta(v)dv} = 0.$$

To get rid of dx/dn we differentiate also the relation (2) between x and n, obtaining an equation exactly like the one above excepting that t is replaced throughout by O. Eliminating dx/dn between the two we have

$$\begin{vmatrix} \int_t^n Y(\tau)e^{-\int_t^\tau \delta(v)dv}\,d\tau & e^{-\int_t^n \delta(v)dv} \\ \int_0^n Y(\tau)e^{-\int_0^\tau \delta(v)dv}\,d\tau & e^{-\int_0^n \delta(v)dv} \end{vmatrix}[xY(n) - O(n) - \delta(n)S(n) + S'(n)] = 0.$$

If the factor in the square brackets vanishes we have (3). But it must vanish, for the determinant becomes, when the upper row is multiplied by $e^{-\int_0^t \delta(v)dv}$,

$$x = \frac{O(n) + \delta(n)S(n) - S'(n)}{Y(n)}, \tag{3}$$

a condition independent of t, as it should be. This equation states that x, the cost of a unit of product is found by adding the operating cost $O(n)$ of the machine (at the time n when it is least efficient and is about to be scrapped) to interest $\delta(n)S(n)$ on the scrap value and the rate of depreciation $-S'(n)$ of the scrap value, and dividing this sum by the machine's rate of production.

If conditions are not static we may still be able to write a known function $x(\tau,\alpha)$ for x in (1) and (2), and $x(n,\alpha)$ for x in (3). We then solve (2) and (3) simultaneously for n and α and substitute in (1). This parallels a method suggested by Dr. Taylor in the paper cited above, but is slightly more general.

There is of course nothing in the theory to preclude the possibility that n may become infinite.

OPERATING EXPENSE DEPENDENT ON VALUE

It has been assumed up to this point that when we set out to solve a particular depreciation problem we know a function $O(t)$ giving the operating expense at all times in the life of the article. But certain elements of operating cost always depend upon the value of the article at the time, which value we do not know until we have solved the problem. Thus taxes are supposed to be proportional to the value, and insurable risk, if not insurance, certainly is. The risk of obsolescence enters here.

If we write $O(\tau)$ in (1) as a function of $V(\tau)$ and τ we have an *integral equation* to solve for the unknown function $V(t)$. Now the study of integral equations is a new and incomplete branch of mathematics, so that an integral equation written down at random can probably not be solved until pure mathematics has advanced further than at present. By rare good luck, however, the cases which usually arise in practice are of a type leading to one of the few integral equations whose theories are well developed, the *Volterra* equation. This is because the dependence of operating cost upon value is ordinarily *linear*. Thus

$$\left| \begin{array}{cc} \int_t^n Y(\tau)e^{-\int_0^\tau \delta(\nu)d\nu}\,d\tau & e^{-\int_0^n \delta(\nu)d\nu} \\ \int_0^n Y(\tau)e^{-\int_0^\tau \delta(\nu)d\nu}\,d\tau & e^{-\int_0^n \delta(\nu)d\nu} \end{array} \right| .$$

which can vanish only if the terms in the first column are equal. This would imply that $\int_0^t Y(\tau)e^{-\int_0^\tau \delta(\nu)d\nu}\,d\tau = 0$, which is impossible (excepting for $t = 0$) because we suppose $Y(\tau)$ positive for all values of τ less than n.

taxes and insurance premiums are directly proportional to the value, and not to its square or some other function. Hence we may write

$$O(\tau) = A(\tau) + B(\tau)V(\tau),$$

where $A(\tau)$ and $B(\tau)$ are functions which, like $Y(\tau)$ and $\delta(\tau)$, are supposed to have been determined, or at least estimated, on the basis of experience. Substituting this value of $O(\tau)$ in (1) we have

$$V(t) = \int_t^n [x \cdot Y(\tau) - A(\tau)] e^{-\int_t^\tau \delta(\nu) d\nu} \delta\tau + S(n) e^{-\int_t^n \delta(\nu) d\nu} \atop - \int_t^n B(\tau) V(\tau) e^{-\int_t^\tau \delta(\nu) d\nu} d\tau. \quad \Bigg\} \quad (4)$$

This integral equation admits of a very easy solution by reduction to a differential equation as follows. Differentiate (4) and add each member of the resulting equation to the product of $-\delta(t)$ by the corresponding member of (4). Several terms cancel out, leaving [1]

$$\frac{dV(t)}{dt} - [\delta(t) + B(t)]V(t) = -xY(t) + A(t). \quad (5)$$

Let us write for brevity $\gamma(t) = \delta(t) + B(t)$. The solution of (5), found by elementary methods, is then

$$V(t) = e^{\int_0^t \gamma(\nu) d\nu} \left\{ \int_t^n [xY(\tau) - A(\tau)] e^{-\int_0^\tau \gamma(\nu) d\nu} d\tau + k \right\}$$

where k is the constant of integration. To evaluate k we may use either the fact that $V(0) = c$ or that $V(n) = S(n)$. In the first case we find

$$V(t) = e^{\int_0^t \gamma(\nu) d\nu} \left\{ c - \int_0^t [xY(\tau) - A(\tau)] e^{-\int_0^\tau \gamma(\nu) d\nu} d\tau \right\};$$

in the second case we obtain the equivalent form

$$V(t) = \int_t^n [xY(\tau) - A(\tau)] e^{-\int_t^\tau \gamma(\nu) d\nu} d\tau + S(n) e^{-\int_t^n \gamma(\nu) \delta\nu}, \quad (1a)$$

which differs from (1) only in that $\gamma(\nu)$ replaces $\delta(\nu)$ and $A(\tau)$ replaces $O(\tau)$.

The problem of value and depreciation is now solved exactly as in the simpler case. Equations similar to (2) and (3) are derived and used in the same manner as before, δ being everywhere replaced by γ and $O(t)$ by $A(t)$.

---

[1] If we solve (5) for x we have the well-known expression for cost of output in terms of operating cost, yield, interest and depreciation on the machine.

EXAMPLE

Suppose that the data available from experience justify the expectation that

$$Y(\tau) = ae^{-\lambda\tau}, \; A(\tau) = be^{\mu\tau}, \; S(n) = 0, \text{ and } \delta(\nu) + B(\nu) = \gamma, \text{ a constant.}$$

Then, by (1a),

$$V(t) = \int_t^n [xae^{-\lambda\tau} - be^{\mu\tau}]e^{-\gamma(\tau-t)}d\tau$$

$$= e^{\gamma t}\int_t^n [xae^{-(\lambda+\gamma)\tau} - be^{(\mu-\gamma)\tau}]d\tau$$

$$= xa\frac{e^{-\lambda t} - e^{\gamma t - (\lambda+\gamma)n}}{\lambda+\gamma} + b\frac{e^{\mu t} - e^{\gamma t + (\mu-\gamma)n}}{\mu-\gamma}.$$

From an equation similar to (3) we have, since $S(n) = 0$ and therefore $S'(n) = 0$, $x = A(n)/Y(n) = be^{(\lambda+\mu)n}/a$.

Substituting above we have

$$V(t) = \frac{b}{(\lambda+\gamma)(\mu-\gamma)}\Big\{(\mu-\gamma)e^{-\lambda t + (\lambda+\mu)n} - (\lambda+\mu)e^{\gamma t + (\mu-\gamma)n} + (\lambda+\gamma)e^{\mu t}\Big\}.$$

For $t = 0$ this gives

$$c = V(0) = \frac{b}{(\lambda+\gamma)(\mu-\gamma)}\Big\{(\mu-\gamma)e^{(\lambda+\mu)n} - (\lambda+\mu)e^{(\mu-\gamma)n} + \lambda + \gamma\Big\},$$

which on rearrangement becomes

$$e^{(\lambda+\mu)n} + \frac{\lambda+\mu}{\gamma-\mu}e^{(\mu-\gamma)n} + \frac{\lambda+\gamma}{\mu-\gamma} - (\lambda+\gamma)\frac{c}{b} = 0.$$

This equation for determining n, it will be observed, contains b and c only in the ratio c/b, and does not contain a at all. Since commensurable values may always be taken for γ, λ and μ it reduces to an algebraic equation. It may be proved mathematically that in every case there is just one root for which $n > 0$, and that n approaches zero and infinity with c/b.

Let us take $\gamma = .09$, $\lambda = .06$, $\mu = .14$, $c/b = 20$. The equation becomes $e^{.20n} - 4e^{.05n} = 0$. Hence $e^{.05n} = \sqrt[3]{4}$, and $n = \frac{1}{3}\log 4/.05 \log e = 9.244$ years. Therefore

$$V(t) = \frac{b}{.15}[e^{-.06t + .20n} - 4e^{.09t + .05n} + 3e^{.14t}]$$

$$= \frac{b}{.15}[4^{4/3}(e^{-.06t} - e^{.09t}) + 3e^{.14t}].$$

Calculation from this formula gives the following results for $b = 1,000$.

t	$V(t)$	$V(t-1) - V(t)$
0	20,000.00	
1	16,553.86	3,446.14
2	13,327.50	3,226.36
3	10,345.07	2,982.43
4	7,638.10	2,706.97
5	5,246.63	2,391.57
6	3,220.65	2,025.98
7	1,621.75	1,598.90
8	524.96	1,096.79
9	20.97	503.99
9.244	0	

The decrease in value shown by this table for the first year is more than $1\frac{1}{2}$ times that indicated by the straight line law, and is more than double that calculated by means of the equal annual payment or sinking fund method with interest at 6 per cent.

Depreciation has been defined as rate of decrease of value:
$D(t) = -dV(t)/dt$.

The *total depreciation* over a period is the difference between the value at the beginning of the period and that at its end. It equals the average value of the depreciation times the length of the period; for

$$\int_a^b D(t)dt = -\int_a^b \frac{dV(t)}{dt}dt = V(a) - V(b).$$

The total depreciation in the value of an article over its whole life, is thus $c - S(n)$. The average depreciation is therefore

$$\frac{c - S(n)}{n}. \tag{6}$$

RELATION TO OTHER DEPRECIATION THEORIES

The formula (6) for average depreciation is so simple that if accuracy is not worth while, and if the useful life n is known with some degree of definiteness, it may be assumed that the depreciation of each year is equal to this average depreciation. Or if we are considering the composite depreciation of a large number of similar machines whose times of installation have been uniformly distributed over n years, (6) gives a rough approximation. This *straight line law* is often used in cases in which it gives rise to large errors.

Let us suppose that $O(t)$ and $Y(t)$ are constants q and y, respectively, for a certain number of years of the machine's life and then change abruptly in such a way as to make it clear that the end has come. The supposition that $O(t)$ is constant means, of course, that $B(t) = O$, that is, that neither risk nor a tax on value exists. Let the scrap value be s, a constant, and let $\delta(t) = \delta$, a constant. Let x also be constant. Then (3) does not hold because its derivation assumed $O(t)$ and $Y(t)$ continuous. But n is now known, so that we can find x from (2) alone.

In this case (1) becomes

$$V(t) = \int_t^n (xy - q) v^{\tau - t} \, d\tau + s v^{n-t}$$
$$= (xy - q) \frac{1 - v^{n-t}}{\delta} + s v^{n-t}.$$

Putting $t = O$ we have the equation, like (2),

$$c = (xy - q) \frac{1 - v^{-n}}{\delta} + s v^n.$$

The result of eliminating $xy - q$ between these two equations is

$$\begin{vmatrix} V(t) - s v^{n-t} & \dfrac{1 - v^{n-t}}{\delta} \\ c - s v^n & \dfrac{1 - v^n}{\delta} \end{vmatrix} = O.$$

Now add $-s\delta$ times the second column to the first. Then subtract the second row from the first. Finally, multiplying the second column by $(1+i)^n$ and remembering that $\bar{s}_{\overline{n}]} = \dfrac{(1+i)^n - 1}{\delta}$, we have

$$\begin{vmatrix} V(t) - c & -\bar{s}_{\overline{t}]} \\ c - s & -\bar{s}_{\overline{n}]} \end{vmatrix} = O.$$

From this we find at once $V(t) = c - (c - s) \dfrac{\bar{s}_{\overline{t}]}}{\bar{s}_{\overline{n}]}} = c - (c - s) \dfrac{s_{\overline{t}]}}{s_{\overline{n}]}}.$

Hence
$$V(t) - V(t+1) = (c - s) \frac{(1+i)^t}{s_{\overline{n}]}}. \tag{7}$$

This gives a proof of the correctness, under the circumstances assumed, of the "equal annual payment" and "sinking fund" methods of calculating depreciation. But these methods are almost always misleading because the assumptions take no account of the almost universal tendency for operating costs to increase and for output to decrease with age. Consequently they understate the depreciation in the early years.

We may further inquire in what circumstances particular depreciation methods are valid. The test here is supplied by (5), which we now write in the form

$$P(t) = \gamma V(t) - \frac{dV(t)}{dt},\tag{8}$$

where $P(t) = xY(t) - A(t)$. Let us, for simplicity, consider γ constant, and let us call $P(t)$ the *gross rental*.

According to the straight line method,

$$V(t) = c - (c-s)t/n.$$

Substituting this expression in (8) we find

$$P(t) = c - s + c\gamma - (c-s)\gamma t/n.$$

This is such a special condition that we must in general discard the straight line method as a valid rule where accuracy is required.

The reducing balance method supposes that depreciation is a constant percentage of value. It follows that $V(t) = ce^{-kt}$, k being determined by $s = ce^{-kn}$. From (8) we have in this case the condition $(P)t = ce^{-kt}(\gamma - k)$.

Similarly the condition for the validity of the sinking fund method (7) is

$$P(t) = \gamma - \frac{c-s}{(1+i)^n - 1}[B(t)(1+i)^t + \gamma],$$

which is constant if $B(t) = O$.

Thus the three methods in most common use depend for their validity upon the satisfaction of conditions which involve no disposable constants and which are so special that the chances are overwhelmingly against the satisfaction of any of them in a particular case.

Dr. Taylor [1] defines unit cost plus by a formula which, in terms of continuous functions, becomes

$$x = \frac{O(t) + \delta(t)V(t) - \dfrac{dV(t)}{dt}}{Y(t)},$$

which is equivalent to (5), and reduces to (3) when $t = n$. We may now derive his criterion of minimum unit cost plus as follows. The value of

[1] J. S. Taylor, *op. cit.*

the machine being given by (1) in terms of x and n, and these unknowns being connected by (2), we seek the value of n which will make x a minimum. This can be found from (2) alone. For, differentiating with respect to n and putting $dx/dn = 0$, we have (3). The simultaneous solution of (2) and (3) will then yield the same values of x and n as those found under (C).

Thus Dr. Taylor's treatment yields the same results as those found under (C), save for the discrepancies arising from the use of discontinuous functions which suppose all changes to occur by yearly steps. It is better adapted to calculation than the theory of the present paper in many cases in which $B(t) = 0$, that is, in which there is no element of operating expense such as risk, insurance and taxes. But even under this severe restriction it does not seem a logical foundation for a theory of depreciation without a demonstration, such as the one above, that the life to be allowed a machine to make it most valuable to its owner is also that which makes unit cost plus a minimum. This proposition is not obvious, and is false in case of the failure of our postulate II.

Dr. Taylor in the same paper mentions as an alternative to that discussed—though without advocating it—a theory based on the assumption that "unit cost," defined like unit cost plus except that interest is not included, should be made a minimum. This assumption lacks even the justification given for minimizing unit cost plus. Like the straight line law, it has nothing to recommend it but simplicity.

FURTHER DEVELOPMENTS

What can be said in case of the failure of postulate II—that the property is used to full capacity? There are very important cases, such as those of mines, in which the postulate is not even approximately true. In this connection we must consider as unknowns not only useful life, value, and depreciation but also the functions $Y(t)$ and $A(t)$. The owner, that is, may voluntarily run the machine at less than full capacity, and wishes us to tell him just how fast to let it run in order that his profits may be a maximum. If we do not tell him, he will guess to the best of his ability. The demand function must be known in order to give a solution.

In all such cases the guiding principle is that the right member of (1), representing discounted future profits, is to be made a maximum. Even if the capitalist system is to give way to one in which service and not profit shall be the object, there will still be an integral of anticipated utilities to be made a maximum. Since we must find a function which maximizes an integral we must in many cases use the Calculus of Variations. But the problem here transcends the questions of depreciation and useful life, and belongs to the dawning economic theory based on considerations of maximum and minimum which bears to the older theories the relations which the Hamiltonian dynamics and the thermodynamics of entropy bear to their predecessors.[1]

The question of charging depreciation as a function of output rather than of time has been discussed of late.[2] It is more natural to consider the depreciation of an automobile in terms of miles than of years. Since interest is an element, this is strictly possible only in case the rate of output $Y(t)$ is known for the whole life of the property; but in this case the arguments commonly advanced for the proposed method fail.

However, an answer may be given to a question of some interest which arises in this connection in case operating cost depends on a controllable rate of output and not merely on the time. What is the additional cost of taking a short trip in an automobile? More generally, what is the additional net cost of a slight increment in output of a machine? We suppose that the increased output extends over a time so short that interest for this period is negligible. The operating cost $A(t)$ per unit of time, exclusive of items such as risk and taxes which are proportional to value, may usually be divided into three classes. One class of costs depends only on the age of the machine, one on age and the amount of current use, and one on the total of past use. We may thus write as an approximation which is probably good enough for all existing data,

$$A(t) = \alpha(t) + a(t)Y(t) + b\int_0^t Y(\mu)d\mu.$$

The additional cost of a small temporary increment z to the total output, concentrated in a time dt, consists partly of increased operating cost at this time and partly of depreciation due to subsequent increase in operating cost. The first part is obviously $za(t)$. The second part depends upon the increment of the integral $\int_0^\tau Y(\mu)d\mu$. This increment will be z for every time τ later than the time t at which the increased use takes place. Thus the future rentals of the machine are decreased by an amount bz per unit of time. The decrease in its value is the discounted sum of these decreases,

[1] For the solution of a special problem in the new "entropy" economics, see G. C. Evans, "The Dynamics of Monopoly," *American Mathematical Monthly* XXXI, 2, February 1924, pp. 77-83. A thorough working knowledge of the Calculus of Variations is a prerequisite to the development of this type of economic theory—which doubtless explains why it has not developed further.

All hedonistic and eudaemonistic ethical theories, which declare that the total of pleasure or happiness should be made a maximum, really reduce the question of right conduct to a set of problems in the Calculus of Variations and in the more general theory of maxima of functionals.

[2] E. A. Saliers, *Depreciation, Principles and Applications*, Ronald Press, 1922, pp. 172-178.

$$\int_t^n bze^{-\gamma(\tau-t)}d\tau = bz\frac{1-e^{-\gamma(n-t)}}{\gamma},$$

if γ is constant.

SUMMARY

(A) The value of a machine and that of a unit of its output are *interrelated*, each affecting the other. This economic truism must underlie a correct theory of depreciation.

(B) The fundamental formula (1) gives the value of a machine in terms of time, value of output, operating cost, scrap value, useful life and rate of interest. One relation between these quantities, given by (2), is due to the fact that the cost of the machine when new is its value at that time.

(C) Two postulates are introduced, one equivalent to the supposition of completely rational action with a completely selfish motive; the second, to which we attach less permanent importance, excludes from consideration the possibility that the owner may seek to increase his profits by slowing down production.

(D) Methods are given for finding the value function $V(t)$ when we know either the useful life n of the machine or the value x of a unit of output, and also when we know neither n nor x.

(E) The hitherto untouched difficulty of operating expenses proportional to value is resolved by means of an integral equation.

(F) A numerical example is worked out which is believed typical of a large class of cases.

(G) The depreciation methods hitherto used are tested and all excepting that of Dr. Taylor are found in general to give false results.

(H) When postulate II is abandoned and the rate of output considered controllable, the methods of this paper are capable of further elaboration to cover a great deal of economic and even ethical theory.

(I) Depreciation cannot, as has been proposed, be charged as a function of output alone, the omnipresence of interest preventing this. The increase of depreciation with output can however be calculated by means of our formulae, given adequate experience tables.

Since this article was written, two papers bearing on the subject have appeared: "Economics and the Calculus of Variations," by G. C. Evans, *Proceedings of the National Academy of Sciences*, Vol. 11, p. 90; and "A Note on the Theory of Depreciation," by J. S. Taylor, *Bulletin of the American Mathematical Society*, Vol. 31, p. 222.

DEPRECIATION, COST ALLOCATION AND INVESTMENT DECISIONS

By HECTOR R. ANTON

The proper measure of the cost of product or services derived from plant investment has long been a matter of controversy among accountants, economists, and businessmen. By and large, the measurement has taken place as part of one or more of three characteristic problems, namely, (1) depreciation, (2) maintenance and repairs, and (3) the actual cost of operating the plant within an operating period. A fourth problem, occasionally introduced—generally as an element of one of the first three—is the treatment of interest on plant investment. This article takes the position that the four problems cannot be considered separately and distinctly, but must be taken together if there are to be made meaningful allocations to periodic income.

BASIC CONCEPTS OF DEPRECIATION

Depreciation has many meanings. According to Bonbright,[1] the meanings may be classified into the following basic concepts of depreciation:

1. *A decrease in value during some defined period.* This meaning is apparently the one most commonly taken in popular usage. It assumes the measureability of value either objectively as the "market value" of the asset or subjectively as its "value to the owner," at two differing dates.

2. *Amortisation cost—the accounting concept.* By this criterion the investment is considered a pre-paid operating expense to be allocated by some systematic method to various fiscal periods over its useful life. In general, this is also the concept adopted for tax purposes.

3. *The appraisal concept of depreciation.* In this method depreciation is considered as the value difference between a new asset seeking to replace the old asset, and the replacement (appraisal) value of the old. The value inferiority of the old asset is the measure of the depreciation.

4. *Impaired serviceability or impaired functional efficiency.* This is not a value or cost concept, but rather an engineering one. It should not be used in any connection in which monetary measurement is important.

Much of the controversy over depreciation allowances has resulted from the existence of these conceptual differences. Specific objectives for which each one of the concepts has been developed can hardly be met by the use of any other

---

[1] Bonbright, J. C., *The Valuation of Property.* McGraw-Hill Book Co., Inc., N.Y., 1937 Ch. X.

concept. Furthermore, all the objectives cannot be met by the use of a single concept. Thus, if the objective of the depreciation allowance is to measure the decrease in value during some defined period, it will be attained through amortisation of cost only by chance. Likewise if the objective is to allocate to periodic revenues some measure of the cost of services obtained from a specific asset, it is unrealistic to assume, or expect, that such "depreciation" will be an adequate measure of the replacement needed.

Inasmuch as the literature on depreciation discusses extensively the relative merits of each of the above enumerated concepts, it is not our intention to pursue this aspect of the problem. Instead, we recognise that the amortisation of cost concept of depreciation is well established in accounting practice.

Fundamentally, cost amortisation follows from the fact that investment in plant is essentially investment in future services to be derived from the plant. The amount of the initial investment is, of course, a factor in the determination of the cost of these future services. The allocation of this cost to periods follows.[2] Technically the cost allocation should bear a direct relationship to the services derived. But, as we have stipulated above, the initial cost of the future services is not the only relevant factor in producing the services. A proper matching of cost and revenue needs to take all cost into consideration. The matching process itself is sound economic and accounting theory.

Accordingly, if we are to concentrate on the amortisation of cost concept of depreciation, a refinement of method and theory is warranted. At the same time, we are the first to admit that the amortisation of cost concept of depreciation does not automatically or necessarily accomplish the objectives of other concepts. It is our intention only to develop meaningful patterns of cost allocation under various given conditions, with the expectation that such formulation will lead to a more general theory of cost allocation.

COST ALLOCATION AND CERTAINTY

The proper measure of the cost of product or services derived from plant investment can be determined by considering all elements of the revenue and outlay streams. By postulating the "certainty" of future events it is possible to derive theoretically, not only the present net worth of an asset,[3] but also the timing of the recovery of its capital cost. Further, the assumption can be made

---

[2] American Accounting Association, "Accounting Concepts and Standards Underlying, Corporate Financial Statements," 1948 Revision. (Reprinted in *Accounting Review*. October, 1948, pages 339-344.) American Institute of Accountants—Accounting Terminology Bulletin, No. 1, page 25.

[3] For a description of this process see Moonitz, M., and Staehling, C. C., *Accounting—An Analysis of its Problems*, Foundation Press, Brooklyn, 1952, Vol. 1, Chapters 5 and 6. Applied to depreciable assets see Canning, J. B., *The Economics of Accountancy*, Ronald Press, New York, 1929, Ch. XIV and Appendix A; Hotelling, H., "A General Mathemetical Theory of Depreciation," *Journal of the American Statistical Association*, Vol. XX, September, 1925, page 340; Roos, C. F., "The Problem of Depreciation in the Calculus of Variations," *Bulletin of the American Mathematical Society*, Vol. XXXIV, March-April, 1928, page 218; Preinrich, G. A. D., "The Theory of Depreciation," *Econometrica*, Vol. VI, July, 1938, page 219,

that management invests in plant assets in order to achieve a specified rate of return. Thus assets would be purchased only when the expectations coincided with the requisite rate of return. Under conditions of certainty, then, the derived value of the asset and *asset cost* would be the same amount. In these conditions the expectations of management would coincide with actuality and this actuality would also govern the initial investment value and asset revaluation by periods.

MANAGERIAL EXPECTATIONS AND COST ALLOCATION

Managerial expectations are reflected in the decision to invest in an asset. The cost of the asset is the quantification of these expectations at the time of purchase. Accountants have adopted cost as the measure of asset value largely because of the objectivity and verifiability of the amount, and because of the "actual investment" by the business. The almost universal adoption of the cost allocation concept of depreciation follows.

Since any method of determining periodic depreciation is either arbitrary or assumes a profit function,[4] the accountant is justified in relating his basic depreciation methods to cost. However, this reasoning is incomplete since, once he has adopted cost allocation, he forgets the rationale for the use of cost. Logic would lead to the conclusion that since the investment decision gives rise to the use of cost, the investment decision also should govern the cost allocation. Managerial expectations would thus constitute the rationale not only for the initial and actual investment, but also for cost allocations to periods. In this way cost allocation would be consistent with the cost basis. Depreciation patterns following these premises are derived later in this article.

COST ALLOCATION AND UNCERTAINTY

When viewed in this light, the cost allocation concept of depreciation provides the accountant's empirical solution to the problem of uncertainty (and to other problems) in the determination of revenue to be allocated to the asset, and the matching of cost to that revenue. That the concept purports to serve as an empirical solution of these problems should not obscure the fact that, given certain assumptions, cost allocation is basically a device used by accountants to deal with the problem of value depreciation. The assumptions in question are very similar to assumptions made in general accounting—the assumption of the concept of the going-concern, and its ramifications in detail, is the most important. Derived value depreciation, and therefore the cost allocation "approximation" of it, must not be confused with changes in market value. Rather, derived value depreciation has to do with "value to the owner" as a reflection of realised earning power. By this approach, cost allocation

---

[4] Preinrich, *op cit.*, page 239, "Outside the limited field of regulated monopoly, the search for the true depreciation method has apparently ended in failure, because the only clue which could be found is the rate of profit itself. This rate is necessarily indeterminate within the life periods of the machines succeeding each other. In any event, therefore, an arbitrary assumption is needed. . . ."

depreciation is to be based on the best available information about revenues and outlays. Only in this sense can complete and realistic *cost allocations* to periodic income be made.[5]

Although direct valuation of fixed assets is impossible by the nature of the joint-procedure, an approximation to it can be made. The gross revenues from a depreciable asset are the result of joint operation of the asset and other segments of the enterprise. The proper allocation of these revenues, however, may be reasonably derived. Empirical studies of cost of operations, repairs, maintenance and so on, are possible, and encouragement of such empirical work can be made a prime objective by accountants.

Our main purpose in this article, to which the rest of it is devoted, is to lay the foundations for a sound approach to the problem of cost allocation.

GENERAL FORMULATION

In conditions of certainty the cost of the asset will approximate to the present value of the future net receipts (or revenues). By definition, the expected return per period is equal to the interest on the investment. It follows that the total charges, including depreciation, in any one period are equal to the difference between the total revenue and the predetermined income. Since current outlay expenses may be deducted from current net receipts, the depreciation charge plus the predetermined income must equal the current net receipts. The depreciation, denoted by D, will in any one year be the difference between the net receipts, denoted by R, and the net income, denoted by iP, where i is the interest rate and P is the present value of future net receipts. As has already been noted, P is equal to the cost of the asset in conditions of certainty.

The quantity P can be expressed thus:[6]

$$(1) \quad P = \frac{R_1}{(1+i)^1} + \frac{R_2}{(1+i)^2} + \frac{R_3}{(1+i)^3} \cdots + \frac{R_n}{(1+i)^n}$$

$$\text{or } (2) \quad P = \sum_{s=1}^{n} \frac{R_s}{(1+i)^s}$$

The amount of depreciation is dependent on the values of P over time and these values are declining as the R stream matures. The charge to be made

---

[5] For example, Terborgh ("*Realistic Depreciation Policy*," Chicago, Machinery and Allied Products Institute, 1954, page 197) contends that the basic depreciation charge should decline over the lifetime of capital goods because of the decline in successive annual service values due to wear and tear and obsolescence (Chapter 4), but supports this contention (Chapter 5) by the use of *market values of second-hand durable equipment*. He thus assumes cost allocation on an opportunity cost of replacement basis. This is untenable unless opportunity of replacement is used merely as an indication of earning power.

[6] For ease of computation, we assume that all payments and all receipts are made at the end of each period. This assumption does not impair the analysis but it does greatly facilitate the solution.

against the asset normally constitutes the credit to the depreciation reserve. The relationship is as follows:

$$(3) \quad D_1 = R_1 - iP$$
$$D_2 = R_2 - i(P - D_1)$$

and (4)
$$D_k = R_k - i(P - \sum_{s=1}^{k-1} D_s)$$

in these equations D_k represents the depreciation over the life of the asset and R_k represents the stream of net revenue over its life.

From formula (4) we can determine a general recursion relation[7] describing how the values of D_k are made up. Thus:

$$(5) \quad D_k + 1 = (1 + i) D_k + (R_k + 1 - R_k)$$

Postulated patterns of either revenue or depreciation can then be studied by the use of this general recursion relation. We might wish to determine the depreciation pattern that, in theory, would emerge from given patterns of net revenue. Alternatively, we might wish to find what was the appropriate revenue pattern, starting with some definite depreciation schedule or formula. Both types of problems are worth solving.

MODELS

Once we have determined the general relationships between the various concepts, we can proceed to study the resultant patterns of depreciation that would follow if there were certain expectations of revenue formulated by the management. It must be remembered that we are here dealing with patterns of cost allocation determined in advance on the basis of expected future revenue. Thus not only is cost one of the determining factors but expectations existing at the time the asset is acquired are also factors helping to determine the allocation of the cost. It is, therefore, with cost allocation rather than with value depreciation that we are here concerned.

It is considered that these patterns of cost allocation are both more realistic and more accurate than cost allocation determined either in arbitary fashion or in a manner that pays attention to only certain of the relevant considerations.[8] We are employing these depreciation patterns within the framework of the cost concept of accounting. The approach does not, however, preclude a similar method for evaluating other items entering into the congregation of income.

---

[7] This recursion relation describes how depreciation varies over time as changes occur in the pattern of revenue.

[8] The several methods by which initial cost has in practice been allocated to revenue in given periods may be classed as follows: (a) Straight-line methods; (b) Production methods; (c) Interest methods; and (d) Decreasing charge methods. In addition there have been advanced various methods of combination costs and replacement. None of the various methods mentioned takes into account all the factors affecting the matching of cost and revenues, as outlined in our approach.

but this extension of the approach will not be discussed in the present article.

In our approach, the depreciation pattern is determined at the same time as the decision is taken to acquire assets that depreciate. The depreciation for any period thus derived is then matched with the actual revenues of that period. Outlay, also, is on an actual basis. If the outcome in any one period is better than was expected, there will be an excess on net income, which will be duly recognised. If the outcome is worse than was expected, then a short fall in net income (perhaps even a loss) will be reported. Thus, there will be no arbitrary smoothing of the curve of profits.

Case I. Constant Net Receipts

In Case I we assume that the revenue from the asset is relatively constant over its economic life: over that period the cost of maintenance, repairs and expenses of operation also are relatively constant. It may be difficult to find such conditions in practice but they may be approximated to if the asset is, for example, a diesel locomotive. If the receipts and outlays are expected to remain constant in perpetuity, depreciation, in the sense of cost allocation, is non-existent; if, however, we make the assumption (which is one that is generally made) that there is a given economic life of the assets, and that at its end the net revenue stream ceases, the conditions for the annuity method are satisfied.

This annuity method of computing depreciation can be described as applying to the situation where $R_1 = R_2 = R_3 = \ldots R_s$. Since the difference $R_{k+1} - R_k = 0$ then, from (5) $D_{k+1} = (1+i) D_k$ and $D_k = (1+i)^{s-1} D_1$. This equation is an exponential one, the value increasing with the increasing k if i is positive. (See Figure 1a.) The exponentially arising depreciation has a time factor i. That is to say, the higher the rate of interest, the greater the increase in depreciation. At the minimum, when i equals 0, depreciation is straight-line. The relevant range is to the point where $D_s = R_s$.

Figure 1a

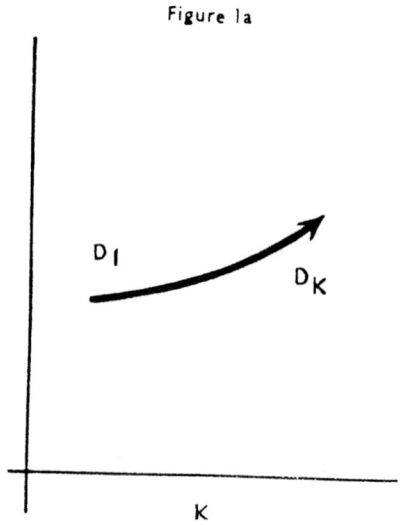

K

In a specific situation of a type covered in class 1, an asset worth $10,000, which has no salvage value and whose life is ten years would, given an interest rate of 8 per cent. per annum, be depreciated as follows:

TABLE I (for Case I)

End of year	Net Receipts $	Net Income at 8 per cent. of Book Value $	Depreciation $	Accrued Reserve for Depreciation $	Book Value of Assets $
0	—	—	—	—	10,000
1	1,490	800	690	690	9,310
2	1,490	745	745	1,435	8,565
3	1,490	685	805	2,240	7,760
4	1,490	621	869	3,109	6,891
5	1,490	551	939	4 048	5,952
6	1,490	476	1,014	5,062	4,938
7	1,490	395	1,095	6,157	3,843
8	1,490	307	1,183	7,340	2,660
9	1,490	213	1,277	8,617	1,383
10	1,490	107	1,383*	10,000	—

*Discrepancy of $4 from rounding-off net receipts.

The relationship may be depicted as in Figure 1b.

It will be seen that by assuming a rate of return—in the example 8 per cent. on the unrecovered investment—the depreciation residual is retained. This process is realistic, given our assumption that by definition the investment is made only if it will yield 8 per cent. per annum. Once the investment is made,

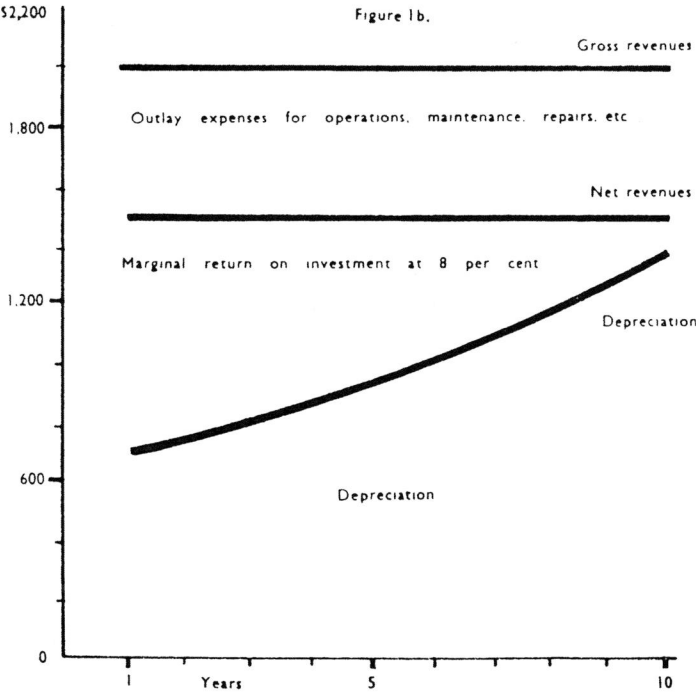

Figure 1b.

the amount of the depreciation follows from the premises and the matching of cost against revenue similarly follows. If actual conditions varied from the model in the example, the variation would be represented by taking a greater or smaller rate of return compared with that expected.

Case II. Net Receipts Decreasing Linearly

In this situation, the assumption is that net revenues are decreasing linearly. Any combination of gross revenue less cost of maintenance, repairs and expenses of operation, that gives this linear decrease in net revenues would fulfil the requirements. However, it is plausible to assume that gross revenue is decreasing linearly and that costs are increasing linearly, although perhaps not at the same rate. But it is the concept of net revenue exclusive of depreciation that is really in question.

Taking our general equation $D_{k+1} = D_k(1+i)+(R_{k+1}-R_k)$, and assuming a linear decrease (a) in R, to obtain $R_k = -a(k-1)+R_1$, we determine that $R_{k+1}-R_k = -a$ and, therefore, that $D_{k+1} = (1+i) D_k - a$. This gives the general solution $D_k = au^k + b$. Substituting in the previous equation, we find that $b = \frac{a}{i}$; $u = (1+i)$ (as before); and thus $D_k = a(1+i)^k + \frac{a}{i}$. Evaluating a, since D_1 and R_1 are known, we find that $D_1 = a(1+i)^1 + \frac{a}{i}$ and a, therefore, is equal to $(D_1-\frac{a}{i}) \frac{1}{(1+i)}$. Dk then is equal to $(D_1-\frac{a}{i}) (1+i)^{k-1}+\frac{a}{i}$. The properties now depend on the co-efficient $D-\frac{a}{i}$, or (D–b) with D and a as given (or determinable), the coefficient sign is determined from the value of i. If i is small, the coefficient is negative and the plotted result is as in Figure 2a.

If $D-\frac{a}{i} =0$, then straight-line depreciation results. If $D-\frac{a}{i}>0$, which might result if i is very large, the above graph of Dk would be reversed, as in Figure 2b.

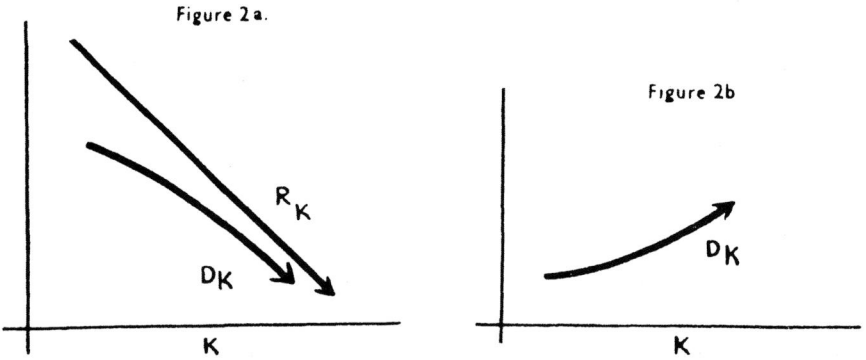

Figure 2a.

Figure 2b

Since the derivation above is based on the recursion formula, it holds over any range in which $R_k - R_{k+1} = a$. This fact is important in considering what happens if the revenue pattern changes. A combination of depreciation patterns as derived here can then be successively depicted.

The value of the asset, its length of economic life and the percentage rate of return are assumed to be as in the previous example. Table II shows the figures.

TABLE II (for Case II)

Year	Net Receipts $	Net Income at 8 per cent. of Book Value $	Depreciation $	Accrued Reserve for Depreciation $	Book Value of Asset $
0	—	—	—	—	10,000
1	2,265	800	1,465	1,465	8,535
2	2,065	683	1,382	2,847	7,153
3	1,865	572	1,293	4,140	5,860
4	1,665	469	1,196	5,336	4,664
5	1,465	373	1,092	6,428	3,572
6	1,265	286	979	7,407	2,593
7	1,065	207	858	8,265	1,735
8	865	139	726	8,991	1,009
9	665	81	584	9,575	425
10	465	34	431	10,006*	—

*Discrepancy of $6 from rounding-off receipts

Relationships as assumed above for case II can now be shown precisely, as in Figure 2c.

Figure 2c

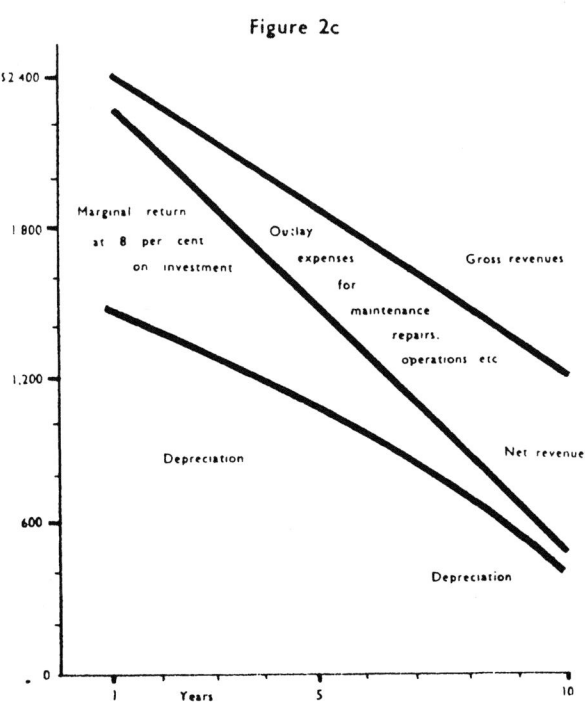

The depreciation will be decreasing at an increasing rate throughout the life of the asset.

Case III. Special straight-line depreciation situation

A special variation of Case II gives straight-line depreciation. This is the situation when $D_1 = \frac{a}{i}$, where a is the periodic decrease in revenue. Thus, with an asset costing \$10,000 as before, to be depreciated over ten years and having no salvage value, and assuming now that the interest rate is 20 per cent. per annum, the resulting schedule of depreciation is a straight-line one, as will be seen from Table III, and depicted in Figure 3.

TABLE III (for Case III)

End of Year	Net Revenue $	Net Income at 20 per cent. of Book Value $	Depreciation $	Accrued Reserve for Depreciation $	Book Value of Asset $
0	—	—	—	—	10,000
1	3,000	2,000	1,000	1,000	9,000
2	2,800	1,800	1,000	2,000	8,000
3	2,600	1,600	1,000	3,000	7,000
4	2,400	1,400	1,000	4,000	6,000
5	2,200	1,200	1,000	5,000	5,000
6	2,000	1,000	1,000	6,000	4,000
7	1,800	800	1,000	7,000	3,000
8	1,600	600	1,000	8,000	2,000
9	1,400	400	1,000	9,000	1,000
10	1,200	200	1,000	10,000	—

Other variants are possible, as long as the decreasing function of revenue is offset by the increasing effect of the interest rate; thus, $D - \frac{a}{i} = 0$.[9]

This example shows what are the necessary conditions for the use of the straight-line method of depreciation. Revenue must decrease rapidly and when the decision to invest is taken, it must be required that the asset should be quickly written-off. In these conditions, depreciation by the straight-line method is more or less accurate! Since the conditions mentioned are conditions of expectations of the management, it may be expected that they obtain not infrequently.[10] To this extent, therefore, it may well be right that straight-line depreciation should be used as widely as it is.

---

[9] In terms of the annuity method of depreciation, this result could be explained as arising when gross depreciation charged to operations decreases by the same amount as the decrease in interest income. But here we make no such distinction. The combined effect is used for purposes of cost allocation only.

[10] Cf. *Cost Behaviour and Price Policy*, National Bureau of Economic Research, pages 162-164.

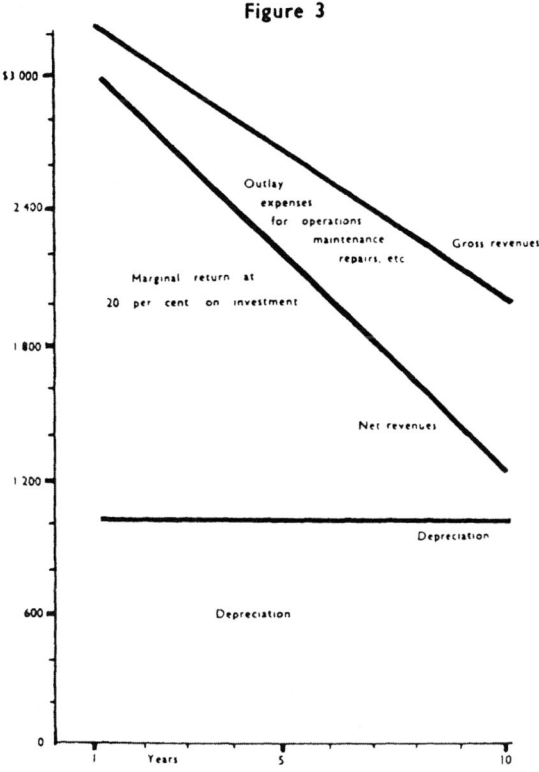

Figure 3

Case IV. Methods of Decreasing Charge Depreciation

Recent legislation in the United States has allowed depreciation for tax purposes on a more liberal basis.[11]

The result of these various alternatives has been greatly to increase the use and importance of the sum-of-the-digits method of computing depreciation. When this method is used, what revenue pattern would make the depreciation pattern theoretically correct? The necessary revenue pattern can be derived generally.

---

[11] The methods that are allowed for tax purposes are set out in the Code rather than in regulations, so removing the matter from administrative decision. In addition to the methods previously allowed, there are now also allowed the following methods: (a) declining balance, with option to convert to straight-line; (b) sum-of-the-digits; (c) other consistent methods. Further, written agreements may be made between the Bureau of Inland Revenue and the taxpayer at the option of the taxpayer as long as the allowances so agreed do not exceed twice those which would be allowed under the straight-line method. (Internal Revenue Code of 1954, Section 167; Public Law 591, Ch. 736, H.R. 8300, 83rd Congress.) Since the sum-of-the-digits method is more advantageous to the taxpayer than the declining balance method, most taxpayers who exercise the option will, it is to be expected, adopt the sum-of-the-digits method.

The sum-of-the-digits depreciation schedule is derived as follows:

$$D_1 = \frac{n}{\sum\limits_{1} k}P; \quad D_2 = \frac{n-1}{\sum\limits_{1} k}P; \quad D_3 = \frac{n-2}{\sum\limits_{1} k}P, \text{ etc.}$$

Since $\sum\limits_{1}^{n} k = n\frac{(n+1)}{2}$,

then $D_k = \dfrac{2(n+1-k)}{n(n+1)} P$ and $D_{k+1} = \dfrac{2(n+1-k-1)P}{n(n+1)}$.

Upon substitution of the recursion relation (5 above), this gives:

$$R_{k+1} - R_k = P\left[2\frac{(n+1-k-1)}{n(n+1)} - (1+i)\frac{2(n+1-k),}{n(n+1)}\right]$$

$$\text{or} = P\left[\frac{2(1+ni+i)}{n(n+1)} - \frac{2ik}{n(n+1)}\right]$$

which is a degenerate solution of a double series. For degenerate cases, the solution is $R_k = A + B_k + Ck^2$. Substituting this into $R_k - (R_{k+1})$ we obtain:

$$\frac{R - Rk+1}{P} = \frac{2(1+in+i)}{n(n+1)} - \frac{2ik}{n(n+1)}$$

$$= A + Bk + Ck^2 - A - Bk - B - Ck^2 - 2Ck - C.$$

Since this method must hold for any value of k, the like powers of k must cancel out separately: Thus for k^2, $C - C = 0$; and for k^1, $B - B - 2C$ $+ \dfrac{2iP}{n(n+i)} = 0$; for k^0, $A - A - B - C - \dfrac{(in+i)P}{n(n+i)} = 0$, and so on.

Substituting:

$$Rk = A - \frac{2+2in+3i}{n(n+1)} Pk + \frac{i}{n(n+1)} Pk^2$$

Shifting co-ordinates and completing the square we obtain:

$$R_k = \frac{iP}{n(n+1)}\left[k - \left(\frac{2+2in+3i}{2i}\right)\right]^2 + A - P\frac{(2+2in+3i)^2}{4n(n+1)i}$$

The more readily recognised form of this equation would be (where $C \sim 0$):

$$y = A(x-a)^2 + B + C$$

Thus, if the depreciation is linearly decreasing, as in the sum-of-the-digits method, the proper revenue pattern is quadrilaterally decreasing (see Figure 4a).

If the sum-of-the-digits method is used, the interest rate being 8 per cent. per annum and the economic life of the asset 10 years, as before, the necessary revenue is shown in Table III. It can be deduced that the revenue is falling quadrilaterally. As before, it should be noted, the quantities n, R, i and P are directly related. If the depreciation pattern is given then n and i are not independent. That is to say, the system would be over-determined if both i and n

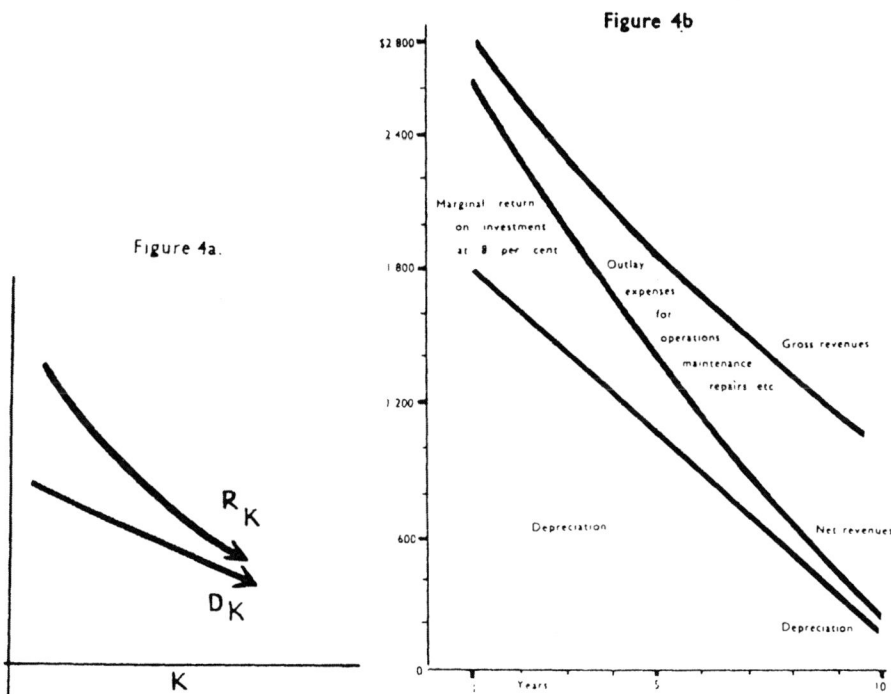

Figure 4a.

Figure 4b

were specified. For this reason, care should be exercised in the use of the formulae: for example, if the sum of the revenues, duly discounted, determines the net cost of the asset, and the depreciation schedule is by the-sum-of-the-digits method and fits the recursion relation, $Dk = R_k - i \left(P - \sum_{s=1}^{k-1} D_s \right)$, then the interest and the number of years in which the asset is written-off are not independent, assuming the net cost of the asset is to be returned during the years over which it is written-off.

The resultant pattern is given in Figure 4b. Table IV shows a revenue pattern decreasing over time but the decrease in revenue is itself also decreasing over time. This gives the same pattern as that which was determined previously.

TABLE IV (for Case IV)

End of Year (k)	Net Revenue $	Net Income at 8 per cent. of Book Value $	Depreciation $	Accrued Reserve for Depreciation $	Book Value of Asset $
0	—	—	—	—	10,000
1	2,620	800	1,820	1,820	8,180
2	2,292	654	1,638	3,458	6,542
3	1,979	523	1,456	4,914	5,086
4	1,681	407	1,274	6,188	3,812
5	1,397	305	1,092	7,280	2,720
6	1,128	218	910	8,190	1,810
7	873	145	728	8,918	1,082
8	633	87	546	9,464	536
9	407	43	364	9,828	182
10	197	15	182	10,000	—

Case V. Combination Patterns

We here assume a pattern of gross revenue that is relatively constant in the early years and then falls rapidly at a linear rate. The cost of maintenance, repairs and expenses of operation are constant in the early years but then increase rapidly. As before, a marginal return on the investment is assumed.

In these circumstances, the pattern of depreciation would be a combination of two patterns already described. Its first part would follow the general pattern for Case I above. In that case the recursion relation, following $R_1 = R_2 = R_3 \ldots = R_s$, gave us $D_k = (1+i)^{s-1} D_1$. Depreciation was increasing exponentially and the rate of its increase was dependent on the value of i. The second part of the pattern, when the revenue is falling, would follow either Case II or Case III. For example, if the revenue started falling in the qth year, it is following the pattern of Case III and we may substitute the formula of that case as from the lth year. Thus:

$$D_k = a(1+i)^k + \frac{a}{i} \text{ but:}$$

$$D_l = a(1+i)^q + \frac{a}{i}, \text{ then:}$$

$$D_k = (D - \frac{a}{i})(1+i)^{k-q} + \frac{a}{i}, \text{ which pattern is as before.}$$

$$k \geqslant q$$

We may, then, take it that if the revenue pattern is constant at first and then falls, the depreciation pattern at first rises exponentially and then falls exponentially (see Figure 5a). This may be seen in the example that follows.

The following illustration shows the depreciation pattern with an asset costing $10,000, as before, having an economic life of ten years, and earning a rate of return of 8 per cent. per annum, again as before. The assumption is now that for the first four years of the life of the asset, the net revenue is constant and that for the remainder of its life, it decreases linearly by $300 per annum. The situation is shown in Table V and the depreciation pattern is depicted in Figure 5b.

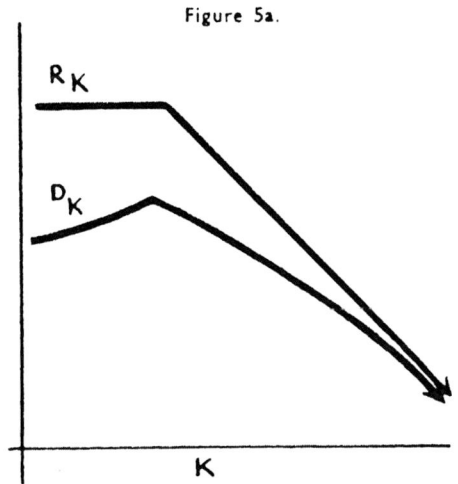

Figure 5a.

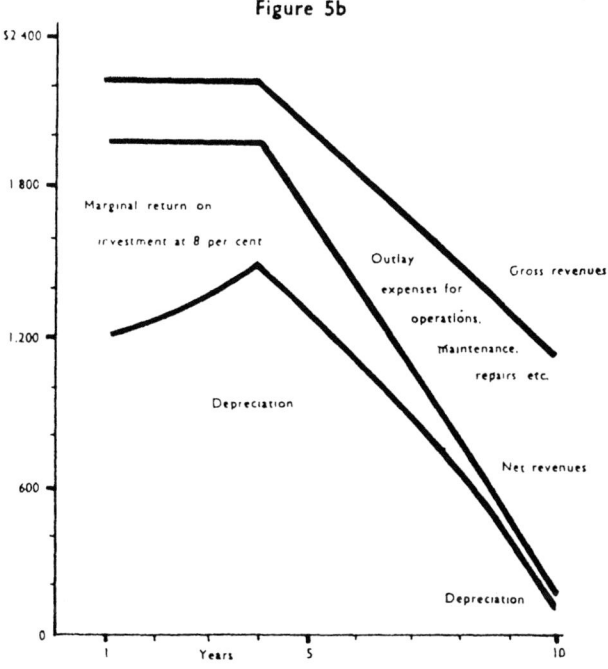

Figure 5b

TABLE V (for Case V)

End of Year	Net Revenue $	Net Income at 8 per cent. of Book Value $	Depreciation $	Accrued Reserve for Depreciation $	Book Value of Asset $
0	—	—	—	—	10,000
1	1,988	800	1,188	1,188	8,812
2	1,988	705	1,283	2,471	7,529
3	1,988	602	1,386	3,857	6,142
4	1,988	491	1,497	5,354	4,646
5	1,688	372	1,316	6,670	3,330
6	1,388	266	1,122	7,792	2,208
7	1,088	177	911	8,703	1,297
8	788	104	684	9,387	613
9	488	49	439	9,826	174
10	188	14	174	10,000	—

Case VI. Increasing, Constant and then Decreasing Revenues

It is usual for revenue to be low at the beginning of the life of an asset but to increase rapidly for a few years, then to remain constant for a time, subsequently decreasing rapidly. The situation might be one, for example, where after installation of an asset time has to elapse before it reaches the stage where it is at its full earning power. We here have three phases—rising revenue, constant revenue and decreasing revenue. In practice, the three stages may well merge into each other in a continuous curve, but for the sake of simplicity we here take them as being related linearly.

Again, the recursion relation is pertinent. We have in the given situation three sections of a curve, in the first of which Dk is rising exponentially at a high rate, in the second of which it is rising exponentially as in the annuity case, and in the last section of which it is either rising exponentially or is constant or is decreasing exponentially, again depending on the co-efficient $D - \frac{a}{i}$. Here, however, a very interesting phenomenon takes place. Since D here is very probably larger than D_1 in a complete straight-line situation (see Case III above), $\frac{a}{i}$ can also be larger. Thus either a (the decrease in revenue) can be larger or i can be smaller and there will still be straight-line depreciation for this segment of the life of the asset. It follows that straight-line depreciation is very likely to be correct, in theory, over an appreciable part of the life. And situations of the type in Case VI are not infrequent. It should be noted, however, that straight-line depreciation is here applied to a segment of the life of the asset following a segment in which depreciation rates were rising; also, the amounts of straight-line depreciation as here derived are substantially higher than the "ordinary" amounts of straight-line depreciation would be.

An example of the depreciation pattern for Case VI is given in Table VI. The same assumptions are made as in the previous case about the cost of the asset, its economic life and the annual rate of return.

TABLE VI (for Case VI)

End of Year	Net Revenue $	Net Income at 8 per cent. of Book Value $	Depreciation $	Accrued Reserve for Depreciation $	Book Value of Asset $
0	—	—	—	—	10,000
1	1,245	800	445	445	9,555
2	1,445	764	681	1,126	8,874
3	1,645	710	935	2,061	7,939
4	1,645	636	1,009	3,070	6,930
5	1,645	554	1,091	4,161	5,839
6	1,645	467	1,178	5,339	4,661
7	1,545	373	1,172	6,511	3,489
8	1,445	279	1,166	7,677	2,323
9	1,345	186	1,159	8,836	1,164
10	1,245	93	1,152	9,988	12

The revenue pattern with the resulting depreciation pattern is depicted in Figure 6.

Figure 6

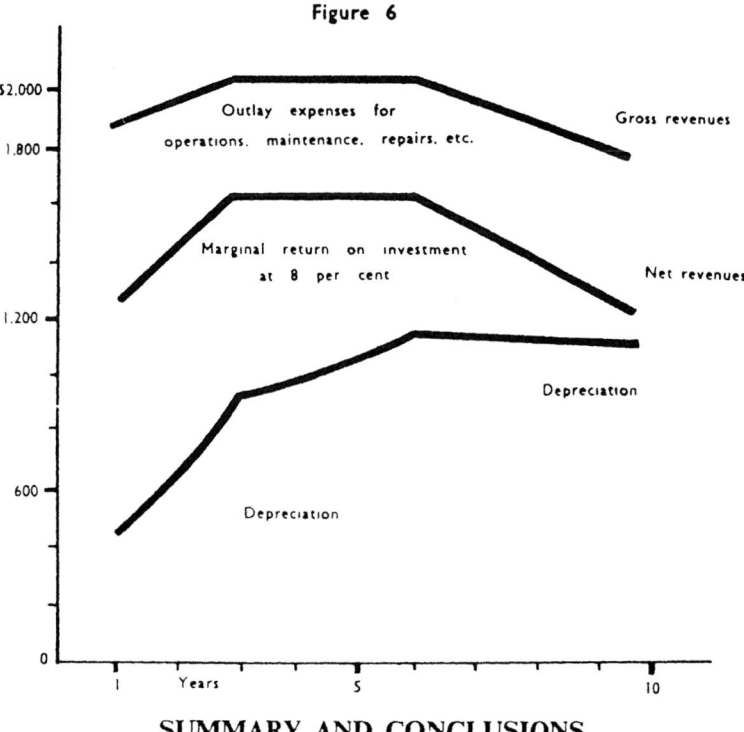

SUMMARY AND CONCLUSIONS

A method has been developed for deriving theoretical patterns of depreciation in various conditions. Perfect certainty as to revenue streams is assumed. The assumption of certainty is carried also to the expectations of revenue streams, which help to determine investment decisions. The "cost principle" in accounting is but one aspect of the situation of this scheme.

It follows that methods of depreciation by "cost allocation" should be based on essentially the same factors as the decision to invest. Since expected expenses and expected revenues determine the cost or at least fix its limits, in an analytical framework the expected expenses and expected revenues must be quantified. Other significant factors are the timing of the expectations and the expected net rate of return on the asset, that is, the rate of interest.

On these premises, it is readily seen that no one method of depreciation by "cost allocation" can apply to all assets or in all circumstances. For example, the straight-line method of depreciation requires rather stringent conditions, namely a high rate of return on the asset and a rapidly declining revenue stream. The theoretical case for various types of "accelerated depreciation" is quite good, especially if the interest rate is low and the revenue stream is rapidly falling. In contrast, all conditions that are usually considered as appropriate for the straight-line method of depreciation give instead increasing depreciation patterns similar to those of the annuity method.

It has also been demonstrated that the level of the rate of interest affects not only the amplitude of the depreciation pattern but also the pattern itself—its configuration. The significance of the interest rate is, therefore magnified.

Given certain expectation of revenue and expenses it is possible to derive, by means of the recursion relation determined in this article, a revenue pattern for the life of an asset. It is to be emphasised that the pattern is one of cost allocation and that the expected values are used only to arrive at such a pattern that will be meaningful. The charges for depreciation have to be matched with actual revenues and expenses in order that periodical income should be determined: thus failure to meet expectations, whether the result is better or worse than expected, is reflected in the periodical income. The depreciation charges are dependent not, like production methods, on actual activity, but rather on activity that was planned when the asset was acquired.

This concept of depreciation gives a formula that is more difficult than the normal one, but its enhanced validity should make its use worth while. Even if the pattern of depreciation is arrived at in approximate fashion, it would be useful in many situations. Further, since in the United States it is now possible for taxpayers and the Government to enter into special agreements on the depreciation allowances that will be granted for income tax purposes, it is possible to work out and to make use of more realistic patterns for assets falling in different groups.

Berkeley, Calif., U.S.A.

Theory of Depreciation: Composite Plant

ANNUAL SURVEY OF ECONOMIC THEORY:
THE THEORY OF DEPRECIATION

By Gabriel A. D. Preinreich

THE LITERATURE of depreciation is very extensive, but it clearly shows that the study of this important subject is being carried on in separate compartments isolated from one another. It is the purpose of the present article to appraise representative samples of independent thought and to expand their valid portions into rudiments of a theory, or a foundation for further development.

Mathematical economists have excogitated more or less fine-spun applications of the law of capital value to so-called capital assets, but most of their attention appears to have been devoted to the economic behavior of a single "machine," rather than to the continuous flow of productive assets through a plant or production center. There has been great interest in why the economic life or usefulness of a machine reaches its end, but comparatively little in the mass phenomena of how gradual consumption actually occurs.

Engineers, on the contrary, emphasize the latter aspect in elaborate mortality studies conducted upon the same principles as for human lives. The results are expressed in time-service units, often considered equivalent to cost units. To describe these units as life units would be more accurate, since the capacity or opportunity for rendering service within successive time units is generally variable. Little or no attention is given to the concept of value, as governed by the economic phenomena of profit and interest. The tendency throughout is to subordinate theory to practice. Thus, for instance, entirely unrelated types of curves are commonly fitted to groups or phases of data which are of the same type, in the sense that one group bears a definite and fixed mathematical relationship to the other.

Accountants are even more "practical-minded" than engineers, and this possibly explains the lack of any notable contribution on their part. In publications, they are usually content to restate what has been said before. The various depreciation methods employed in practice are always described in the same terms, with hardly any attempt at scientific analysis. Nevertheless, some of the truisms and generalities recopied and re-expounded from year to year have more merit than economists are willing to concede.

Business men and management experts have also contributed millions of words to the theory of depreciation, it being quite apparent in many cases that the wish is the father of the thought. This is true particularly in the public-utility field, where the battle still rages with

219

unabated vigor on the question of whether or not a used machine can be worth less than it cost, so long as it serves as well as a new one.

I

To emphasize that the stream of machines gradually consumed and replaced is more important than its component elements, it seems best to begin with a review of engineering studies of mortality. Dr. Edwin B. Kurtz[1] has compiled a wealth of actual data on 52 different kinds of equipment, all of which show a striking similarity of behavior.

When a large number of similar items of equipment is installed at the same time, the rate at which they will drop out of service forms a bell-shaped, but usually skew, *frequency distribution*. Let it be drawn in such a form that its ordinates $f(0)$ at the time $t=0$ and $f(n)$ at the time $t=n$ be zero, and that the total area enclosed by the curve and the axis of abscissas be equal to unity. For any other time t the curve will be denoted by $f(t)$. Summation of $f(t)$ from t to n gives the *mortality* curve per centum of the total number of machines installed at the time $t=0$:

$$(1) \qquad M(t) = \int_t^n f(\tau)d\tau; \qquad M(0) = 1; \qquad M(n) = 0.$$

The area enclosed by the mortality curve and the co-ordinate axes represents the total life units of service expected from the machines installed at the outset. Since the ordinates are expressed per centum, the area also equals the *average life* of the machines, i.e., their life expectancy at $t=0$:

$$(2) \qquad a = \int_0^n M(\tau)d\tau.$$

The *life expectancy* of the machines in service at any other time t is the quotient of the remaining life units of service by the mortality curve:

$$(3) \qquad L(t) = \frac{1}{M(t)} \int_t^n M(\tau)d\tau.$$

The sum of the age t and the life expectancy $L(t)$ is the *probable life* of any machine still in service. If the rate of service rendered by each machine is contant, the remaining life units of service, expressed per centum of the original units, may be called the *remaining services per centum*. For a single machine, the information is obtained by dividing

---

[1] *Life Expectancy of Physical Property*, New York, Ronald Press Co., 1930.

the life expectancy by the probable life. In terms of all machines originally installed, the formula is

$$(4) \qquad R(t) = \frac{1}{a} \int_t^n M(\tau)d\tau.$$

From the analysis of the life characteristics of the machines, it is possible to compute the *rate of renewals* required to keep the number of machines in service constant. The rate of replacement of the original machines is evidently the basic frequency distribution. In addition, however, the replacements must also be replaced, it being assumed that the new machines will behave in the same way as the original ones.

Let us divide the area of the frequency distribution $f(t)$ and that of the mortality curve $M(t)$ into vertical strips of equal width Δt, beginning at the origin and proceeding toward the right. If k be the total number of strips, $t = k\Delta t$. The area of the first strip $f_1\Delta t$ represents the number of machines (per centum) which must be left at the end of the period Δt_1, out of the unknown number $u_1\Delta t$ actually added during Δt_1. Similarly $f_1\Delta t + f_2\Delta t$ is the number which must be left at the end of $2\Delta t$, out of the unknown total $u_1\Delta t + u_2\Delta t$ added during $2\Delta t$, and so on up to $k\Delta t$. But the average height of each successive strip of the mortality curve, taken from t backward, expresses that proportion of each successive addition, which is still in service at the time t. Thus, $M_k u_1\Delta t$ is left of the machines installed during Δt_1, $M_{k-1}u_2\Delta t$ of those installed during Δt_2, and so forth. Therefore

$$(5) \qquad \sum_{i=1}^k f_i\Delta t = \sum_{i=1}^k M_{k+1-i}u_i\Delta t, \qquad i = 1, 2, 3, \cdots, k$$

and

$$(6) \qquad u_k\Delta t = \frac{\sum_{i=1}^k f_i\Delta t - \sum_{i=1}^{k-1} M_{k+1-i}u_i\Delta t}{M_1}.$$

By successively placing $k = 1, 2, 3, \cdots$, the total renewal curve may be constructed step by step. That is no doubt the manner in which Dr. Kurtz has computed his table 50 for the total annual renewals of property group VII.[2]

The *equation of the renewal curve* does not appear to have been developed by anyone up to the present. It can be found by taking the limit of (6). For $\Delta t = dt = 0$, $M_1 = 1$, and

$$(7) \qquad 0 = \int_0^t f(\tau)d\tau - \int_0^t u(\tau)M(t - \tau)d\tau.$$

[2] His own explanation is somewhat more complicated. Cf. *op. cit.*, pp. 168–176.

This is a *Volterra equation of the first kind.* Its derivative,

$$(8) \qquad u(t) = f(t) + \int_0^t u(\tau)f(t - \tau)d\tau,$$

is known as a *Volterra equation of the second kind.* The literature of the latter is extensive, although it is largely restricted to linear forms. No specific reference to frequency distributions is made anywhere; nevertheless some of the methods discovered are applicable.[3] Limiting comment to the present problem, it may be said that, whenever it is possible to transform (8) into a differential equation of finite order, that seems the best way of solution. This can often be accomplished either by differentiating until the kernel $f(t-\tau)$ vanishes or—if it is indefinitely differentiable—until previous derivatives can be so transformed or combined that the integral will be the same as in the last derivative. If one equation is then subtracted from the other, the integral vanishes.

When this method fails, a solution in the form of an infinite series is still possible by expanding both sides, performing the integration term by term, and equating the coefficients of equal powers of t. That is really what most of the other methods amount to, even in cases where a differential equation of finite order could have been obtained. They merely afford various short cuts and, in simple cases, lead to the summation of the resultant series.

Dr. Kurtz has computed the value of the Pearsonian criterion K for each of his seven property groups and finds it negative throughout.[4] From this he concludes that the type of frequency distribution to be fitted to his data is in all cases the Pearsonian curve I, viz.:

$$(9) \qquad y = y_0 \left(1 + \frac{t}{a_1}\right)^{m_1} \left(1 - \frac{t}{a_2}\right)^{m_2}.$$

Differences in the behavior of the seven property groups are expressed by variations in the values of the constants. By way of illustration, let us place $m_1 = 1$, $m_2 = 2$, and $a_1 + a_2 = n$. The result closely reflects the behavior of group VII, which happens to be the largest group, embracing 17 different types of industrial equipment out of the 52 examined.

When the start of the curve is transposed to the origin and the area of the bell is reduced to unity, equation (9) takes the form

$$(10) \qquad f(t) = \frac{12}{n^4} t(n - t)^2.$$

---

[3] For a survey of these methods and an exhaustive list of references see: *Indiana University Studies* by H. T. Davis, especially Nos. 88–90.

[4] *Op. cit.,* p. 91.

If the frequency distribution (10) is substituted in the general formula (S), four successive differentiations lead to a differential equation of the fourth order, the solution of which can be written at sight:

(11) $u(t) = Ae^{\alpha t} + Be^{\beta t} + e^{\gamma t}(C \cos \delta t + D \sin \delta t).$

The roots α, β, and $\gamma \pm \delta i$ are found by solving the quartic auxiliary equation, while the constants A, B, C, and D are obtainable from four simultaneous linear equations expressing relationships found in the process of differentiating both (8) and (11). It should be noted that equation (11) is valid only for the first life cycle, $0 \le t \le n$, because the curve $f(t)$ beyond the bell must be eliminated. For each subsequent cycle a new equation is needed in the general form

(12) $u_{j+1}(t) = \int_{t-n}^{nj} u_j(\tau)f(t - \tau)d\tau + \int_{nj}^{t} u_{j+1}(\tau)f(t - \tau)d\tau,$
$$j = 1, 2, 3, \cdots ;$$

where $u_j(t)$ is known from the previous calculation. It is unnecessary to continue beyond $j=2$, because seven-place logarithms soon become unable to distinguish the result from the straight line $1/a$. The solution of $u_j(t)$ is similar to (11), except that the coefficients are not constants, but polynomials of degree $j-1$ in t.

The renewal curve will consist of separate equations pieced together, whenever the bell of the frequency distribution has a finite range. Where the range is infinite, there will be only a single equation, valid until $t = \infty$. In either case, the general form of the curve resembles the damped oscillations of the governors of steam turbines, etc., when they seek a new equilibrium. The resemblance is closer when the bell has an infinite range, because it is then essential that the real part of the complex roots of the auxiliary equation be negative. Damping can not occur otherwise.[5] When the frequency distribution (10) is employed, the real part of the complex roots is positive and damping takes place merely because the equation is changed. This may be of interest to Dr. Kurtz, who considers the renewal curve as a simple damped sine function, as soon as it crosses its axis $1/a$ for the first time. Judging by his own data,[6] such a curve appears to fit none too well, especially for property group VII. When the real equation of the renewal curve is very complicated, (6) furnishes a more satisfactory approximation, provided Δt be taken small enough. Another simplification, which Dr. Kurtz adopts, is the use of the Gompertz-Makeham formula in lieu of

---

[5] For requirements in the construction of governors, which have a general bearing on the renewal curve, see W. Hort, *Technische Schwingungslehre*, Berlin, Springer, 1922, pp. 266 *et seq.*

[6] Kurtz, *op. cit.*, pp. 178–181.

the actual mortality curve (1). Such steps are often necessary in practice, but it is also worth bearing in mind what the true relationships are.

II

Attempts to compare various methods of depreciation in terms of a composite plant were greatly hampered up to the present by the lack of a simple general expression for the rate of renewals. With the aid a continuous renewal function[7] the task becomes easy.

To make the conclusions as general as possible, *expansion and increasing replacement costs* will also be introduced. It follows immediately from equation (7) that, if the plant is not only maintained, but also expanded at any variable rate $x(t)$, the curve of replacements and additions is obtainable from the relation

$$(13) \qquad \int_0^t f(\tau)d\tau + e^{\int_0^t x(\tau)d\tau} - 1 = \int_0^t u(x, \tau)M(t - \tau)d\tau;$$

$$(14) \qquad u(x, t) = f(t) + x(t)e^{\int_0^t x(\tau)d\tau} + \int_0^t u(x, \tau)f(t - \tau)d\tau.$$

When the price of new machines increases at any variable rate $q(t)$, the renewal function in terms of money may be written for brevity

$$(15) \qquad u(x, q, t) = u(x, t)e^{\int_0^t q(\tau)d\tau}.$$

No matter what method of depreciation be employed, the *book value* $C(x, q, t)$ *of a composite plant* consists of the book value $c(t)$ of the original machines and that of all renewals and additions:

$$(16) \qquad C(x, q, t) = c(t) + \int_0^t u(x, q, \tau)c(t - \tau)d\tau \qquad \text{for } t < n$$

and

$$(17) \qquad C(x, q, t) = \int_{t-n}^t u(x, q, \tau)c(t - \tau)d\tau \qquad \text{for } t > n.$$

The book value $c(t)$ includes both services not yet rendered and the scrap value. If the scrap value of a single machine be $s(t)$, that of all original machines in service at the time t is $\int_t^n s(\tau)f(\tau)d\tau$. These machines depreciate at the rate of change of the difference between the book value and the scrap value. An operation similar to that indicated in (16) leads to the result:

---

[7] For purposes of general discussion, the symbol u will represent the renewal function for the entire interval $0 \leqq t \leqq \infty$, irrespective of how many separate equations it may actually consist of.

(18) $D(x, q, t) = C'(x, q, t) - S'(x, q, t) - [1 - \int_0^n s(\tau)f(\tau)d\tau]u(x, q, t).$

The *composite rate of depreciation* is obtained by finding the rate of change of the composite book value and deleting from the result both the rate of change of the composite scrap value and the renewals of the services consumed.

The best-known methods of depreciation may now be compared by developing the book-value curve $c(t)$ of the original machines, as defined by each. The methods selected are: 1. the straight-line method; 2. the annuity method (equivalent to the sinking-fund method when only one rate of interest is used); 3. the diminishing-balance method; and 4. the retirement method (here called a depreciation method purely for convenience of language). All methods assume that the scrap value of a single machine is a constant proportion of its purchase price.

The *straight-line method* is directly connected with mortality theory in the sense that it counts the unexpired life units and considers them equivalent to the service or value units to be recovered. To these units the scrap value must be added. The book value of many machines in-installed at the same time is therefore by (4):

(19) $c_s(t) = \int_t^n \left[\frac{b}{a} M(\tau) + sf(\tau) \right] d\tau; \qquad b = 1 - s.$

The *annuity method* makes the same assumptions, but it also discounts the value of both future services and scrap proceeds at some rate of return i. This procedure changes (19) to:

(20) $c_a(t) = \int_t^n [\phi M(\tau) + sf(\tau)]e^{i(t-\tau)}d\tau; \quad \phi = \dfrac{1 - s \int_0^n f(\tau)e^{-i\tau}d\tau}{\int_0^n M(\tau)e^{-i\tau}d\tau}.$

The *diminishing-balance method* is a crude attempt to recognize a decline in the value of services, as a machine grows older. Instead of applying the constant depreciation rate b/a to the original cost of the machines still in service, it applies the constant rate $k = \frac{1}{a} \log_e s$ to the diminishing book value:

(21) $c_d(t) = -\dfrac{1}{k} [c_d'(t) + sf(t)] = e^{-kt} - s \int_0^t f(\tau)e^{k(\tau-t)}d\tau.$

The *retirement method* is based upon the false claim that a machine which serves as well as a new one can not be worth less than it cost.

Depreciation is disregarded altogether, until the machine is scrapped. At that time the entire difference between cost and scrap value becomes an operating expense. The book-value curve of the machines originally installed accordingly equals the mortality curve:

$$(22) \qquad\qquad c_r(t) = M(t).$$

Formulae (19) to (22) must now be inserted in the general expression (16) to find the composite book value for all four methods. Special cases lead to simple answers. For instance when replacement costs are static and expansion proceeds at a constant rate, the derivative of (16) is, for the straight-line method:

$$(23) \qquad C_s'(x, t) = - \frac{b}{a} \epsilon^{xt} - s[u(x, t) - x\epsilon^{xt}] + u(x, t).$$

The reduction is made possible by the relations (13) and (14). The simplified composite-book-value formula is then:

$$(24) \qquad C_s(x, t) = 1 + \int_0^t \left[bu(x, \tau) + \left(sx - \frac{b}{a} \right) \epsilon^{x\tau} \right] d\tau.$$

Proceeding similarly for the other methods, we obtain

$$(25) \qquad C_a(x, t) = e^{it} + \int_0^t [bu(x, \tau) + (sx - \phi)\epsilon^{x\tau}]e^{i((t-\tau)}d\tau,$$

$$(26) \qquad C_d(x, t) = e^{-kt} + \int_0^t [bu(x, \tau) + sx\epsilon^{x\tau}]\epsilon^{k(\tau-t)}d\tau$$

$$(27) \qquad C_r(x, t) = e^{xt}.$$

The corresponding depreciation curves are by (18):

$$(28) \qquad\qquad D_s(x, t) = - \frac{b}{a} e^{xt},$$

$$(29) \qquad\qquad D_a(x, t) = - \phi e^{xt} + iC_a(x, t),$$

$$(30) \qquad\qquad D_d(x, t) = - kC_d(x, t),$$

$$(31) \qquad\qquad D_r(x, t) = - b[u(x, t) - x\epsilon^{xt}].$$

So far all curves are given in terms of the original cost of the plant, which was the unit chosen for the purpose. Further study is facilitated by expressing results in terms of the expanding cost of all machines in service at any time t. This base is the book value (27) obtained by the retirement method. If all formulae are divided by e^{xt}, the curves will in due course become horizontal lines. Their respective levels are of interest. To find these ultimate levels, that of the rate of replacements and additions must be found first. For $t = \infty$ and $x(t) = x$, equation (13) may be written:

$$e^{xx} = u(x, \infty)e^{-xx} \int_{\infty-n}^{\infty} M(\infty - \tau)e^{x\tau}d\tau = u(x, \infty) \int_{0}^{n} M(\tau)e^{-x\tau}d\tau,$$

$$(32) \qquad u(x, \infty)e^{-x\infty} = \frac{1}{\displaystyle\int_{0}^{n} M(\tau)e^{-x\tau}d\tau} = \chi.$$

The ultimate discounted levels of curves (24) to (31) are now given by the necessary condition:

$$(33) \quad \underset{t=\infty}{\text{limit}} \ \frac{d}{dt} \ [C(x, t)e^{-xt}] = 0; \quad C(x, \infty)e^{-x\infty} = \frac{1}{x} C'(x, \infty)e^{-x\infty}.$$

Therefore

$$(34) \qquad C_s(x, \infty)e^{-x\infty} = \frac{1}{x} \ \left(bx + sx - \frac{b}{a}\right),$$

$$(35) \qquad C_a(x, \infty)e^{-x\infty} = \frac{1}{x - i} (bx + sx - \phi),$$

$$(36) \qquad C_d(x, \infty)e^{-x\infty} = \frac{1}{x + k} (bx + sx),$$

$$(37) \qquad C_r(x, \infty)e^{-xx} = 1,$$

$$(38) \qquad D_s(x, \infty)e^{-x\infty} = - \frac{b}{a},$$

$$(39) \qquad D_a(x, \infty)e^{-x\infty} = - \phi + iC_a(x, \infty)e^{-x\infty},$$

$$(40) \qquad D_d(x, \infty)e^{-x\infty} = - kC_d(x, \infty)e^{-x\infty},$$

$$(41) \qquad D_r(x, \infty)e^{-x\infty} = - b[\chi - x].$$

Formulae for entirely static conditions are readily obtained by placing $x = 0$, in which case $\chi = 1/a$. A slight difficulty arises only for the straight-line method, since equation (34) becomes indeterminate and must be evaluated.

$$(42) \quad C_s(\infty) = \frac{0}{0} = \underset{i=0}{\text{limit}} \ C_a(\infty) = s + \frac{b}{a^2}\int_{0}^{n} \int_{\tau}^{n} M(\nu)d\nu d\tau.$$

When replacement costs are changing, the general equation (16) can not be simplified to any considerable extent. A simple case was pre-

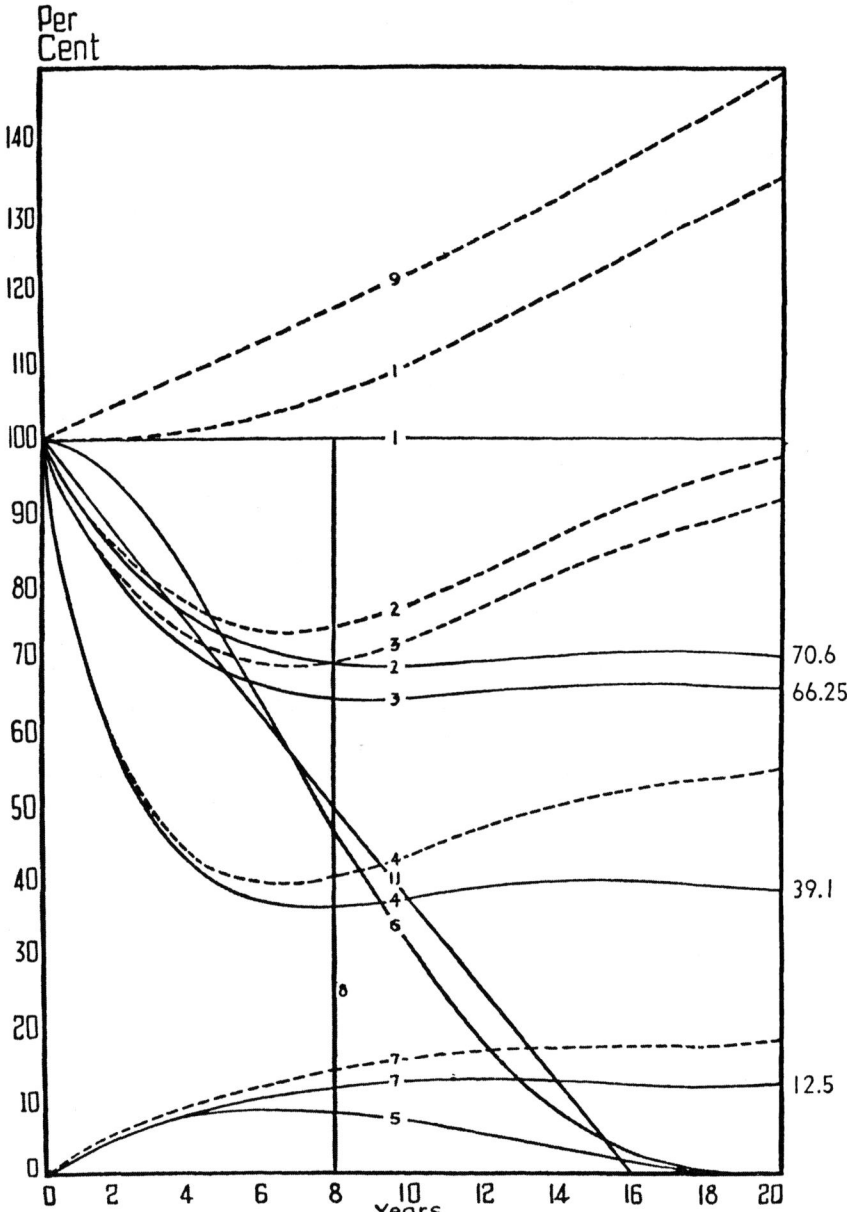

FIGURE 1.—Vertical scale represents per cent of cost of original plant. Smooth lines indicate static conditions. Broken lines include increase in replacement costs at the force of 2 per cent per annum.

Composite-book-value curves: (1) Retirement method; (2) Annuity method at the force of 7 per cent per annum; (3) Straight-line method; (4) Diminishing-balance method.

Auxiliary curves: (5) Basic frequency distribution; (6) Mortality curve; (7) Rate of replacements; (8) Average life = life expectancy of a new machine; (9) Index of replacement costs; (11) Limit of shape of possible other mortality curves. The opposite limit is (8).

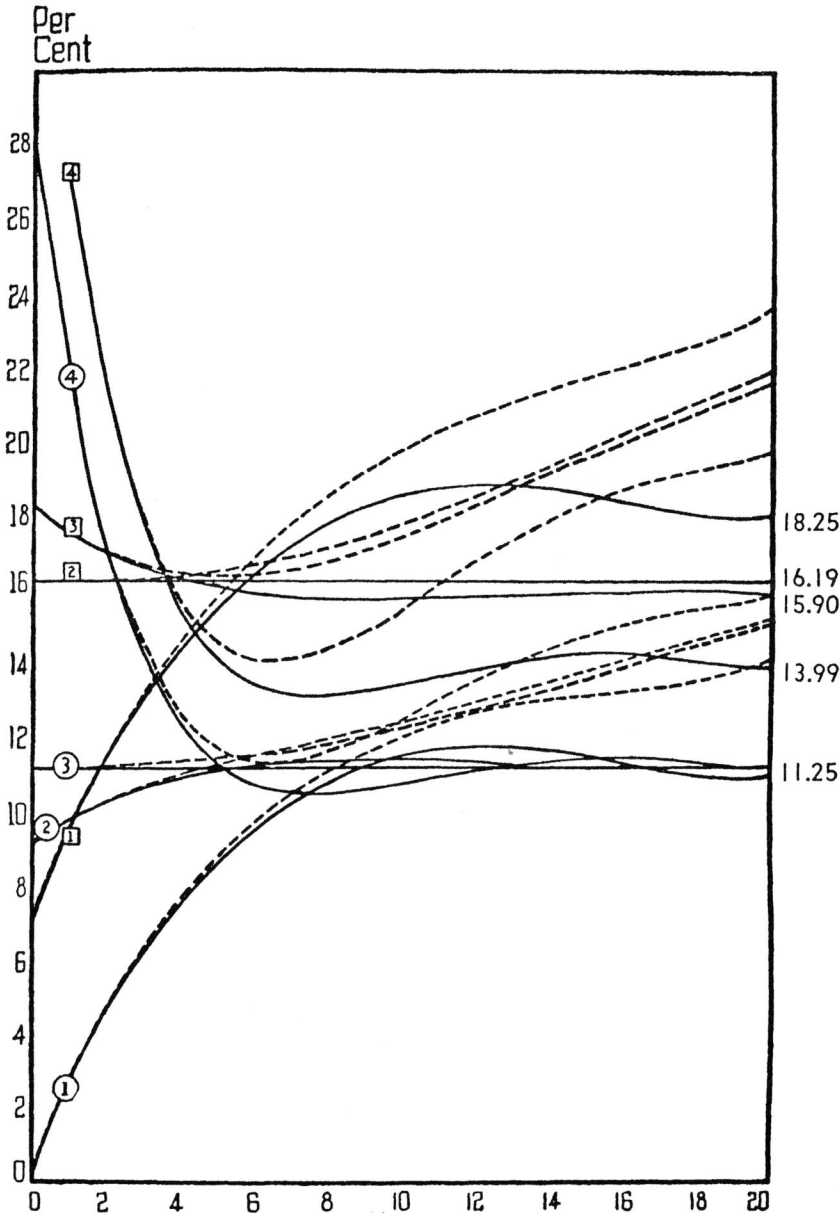

FIGURE 2.—Vertical scale represents per cent per annum of cost of original plant. Smooth lines indicate static conditions. Broken lines include increase in replacement costs at the force of 2 per cent per annum. Circles around numbers indicate rates of depreciation per annum. Squares around numbers indicate the sum of depreciation rates and a fair return at the force of 7 per cent per annum on the respective book values.

1. Retirement method; 2. Annuity method at the force of 7 per cent per annum; 3. Straight-line method; 4. Diminishing-balance method.

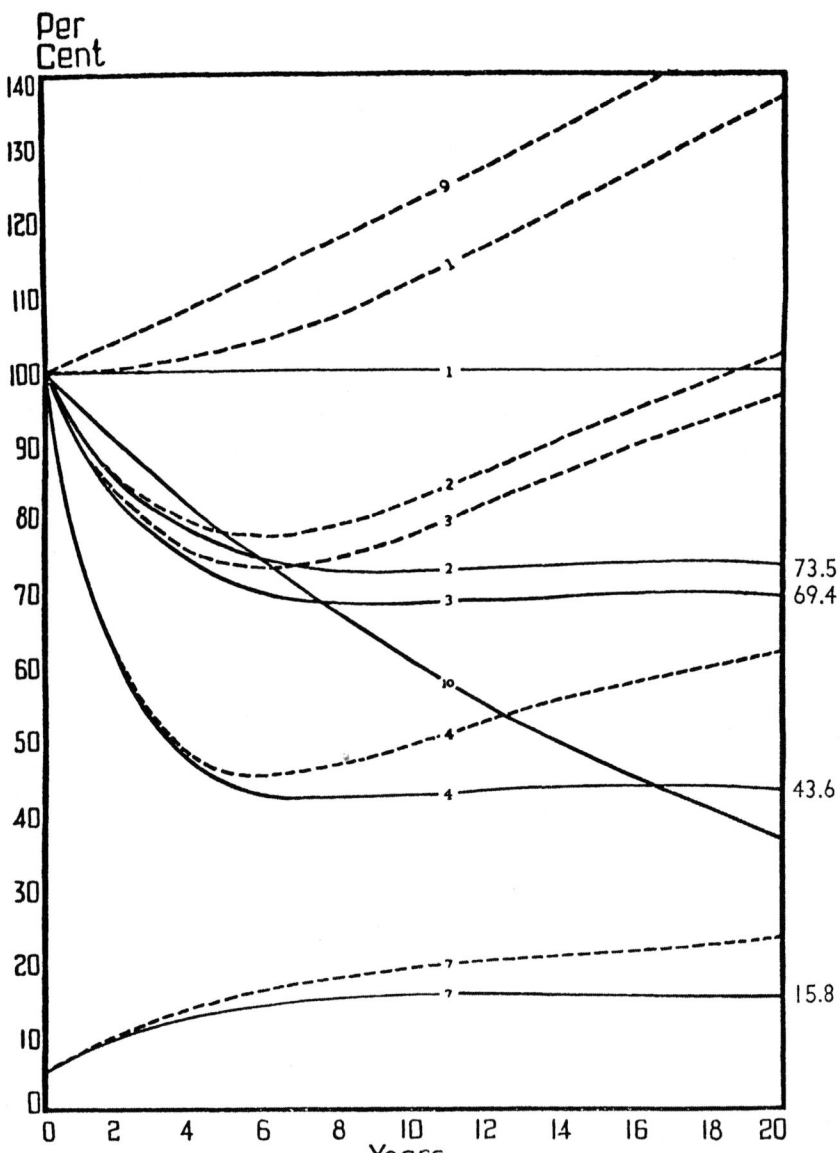

FIGURE 3.—Vertical scale represents per cent of cost of plant in service at any time *t*, when number of machines expands at the force of 5 per cent per annum and when replacement costs remain at the level of original costs. Smooth lines indicate static replacement costs and plant expansion at the force of 5 per cent per annum. Broken lines include increase in replacement costs at the force of 2 per cent per annum.

Composite-book-value curves: (1) Retirement method; (2) Annuity method at the force of 7 per cent per annum; (3) Straight-line method; (4) Diminishing-balance method.

Auxiliary curves: (7) Rate of replacements and additions; (9) Combined index of expansion and replacement costs; (10) Original cost of plant. True ordinates of all curves have been discounted at the force of 5 per cent per annum before plotting.

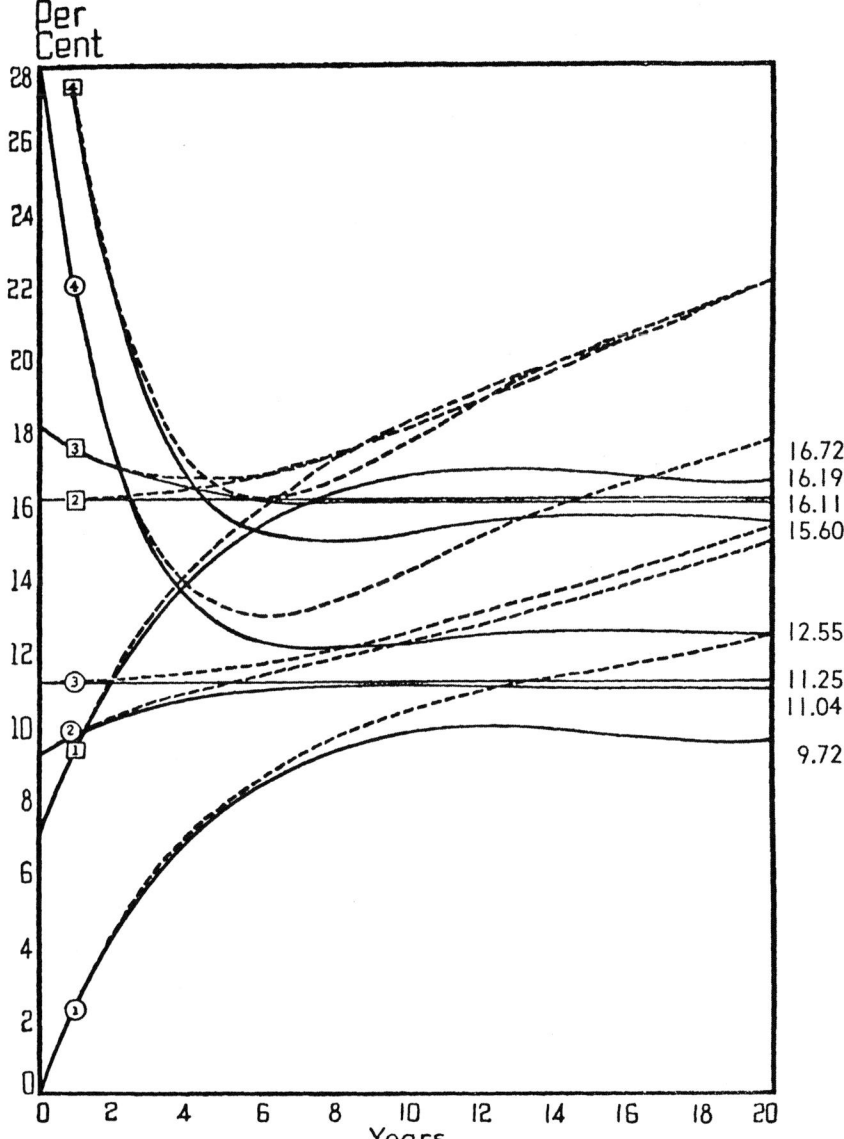

FIGURE 4.—Vertical scale represents per cent per annum of cost of original plant, chargeable at any time t per the original number of machines. Smooth lines indicate static replacement costs and plant expansion at the force of 5 per cent per annum. Broken lines include increase in replacement costs at the force of 2 per cent per annum. Circles around numbers indicate depreciation rates per annum. Squares around numbers indicate the sum of depreciation rates and a fair return at the force of 7 per cent per annum on the respective book values.

1. Retirement method; 2. Annuity method at the force of 7 per cent per annum; 3. Straight-line method; 4. Diminishing-balance method.

sented merely to outline the technique; book-value curves and depreciation rates can be compared far more effectively in graphic form. The four graphs submitted herewith illustrate the following cases:

A. Static conditions.

B. The number of machines in service remains constant. Replacement costs increase at the force of 2 per cent per annum.

C. The number of machines in service increases at the force of 5 per cent per annum. Replacement costs are static.

D. The number of machines in service increases at the force of 5 per cent per annum. Replacement costs increase at the force of 2 per cent per annum.

The scrap value is 10 per cent of the purchase price of a machine; the maximal age n is twenty years. The annuity method was calculated at the force of 7 per cent per annum. These assumptions still simplify actual conditions greatly, but they lead nevertheless to a number of important conclusions:

1. For any method, except the retirement method, the book value of an expanding plant stabilizes itself at a higher proportion of cost than that of a static plant. An expanding plant contains a higher proportion of relatively new machines; its average life expectancy is greater.

2. In the entirely static case, any method of depreciation will ultimately produce the same charge to operations, i.e., a charge at the rate of renewal b/a of the services consumed. The amount of profit reported by the books will ultimately be independent of the depreciation method. The same profit, however, will be measured per centum of widely differing bases.

3. When expansion is present, differences in depreciation charges will never disappear. Books kept by the retirement method will report the lowest charge and therefore the highest profit, while the diminishing-balance method leads to the opposite result. Simultaneously the highest profit will appear to be the lowest one, per centum of the investment reported, and vice versa.

4. When expansion proceeds at a constant rate, the difference between the amount of profit and the amount of expansion (reinvestment) reported by the books is ultimately the same, no matter what method of depreciation be employed.[8] This means that dividend policies ultimately become independent of the depreciation method. The same is true in point 2 above because, in the static case, the entire profit is available in cash.

5. In terms of the number of machines in service, the depreciation

[8] Formulae (34) to (41) show that $xC(x, \infty)e^{-x\infty} - D(x, \infty)e^{-x\infty} = b\chi + sx$ for any method.

rate of the straight-line method remains unaffected by expansion. In the case of the annuity method, the sum of depreciation and a fair return shows similar independence of both expansion and the age of the plant.

6. Until the renewal rate is stabilized, increasing replacement costs increase book values relatively faster for an expanding plant than for a static one. The proportionate difference between depreciated present reproduction cost (appraisal) and depreciated historical cost is smaller for an expanding plant than for a static one. This development occurs in addition to that described in point 1 above.

7. In terms of the number of machines in service, increasing replacement costs increase depreciation charges under any method, but discrepancies are also created in the same sense as by expansion.

8. Both expansion and increasing replacement costs speed the automatic stabilizing process, i.e., they decrease the relative amplitudes of oscillations from the start.

The graphs also present the case of regulated monopolies for the assumption that the force of 7 per cent per annum is a fair return on the "rate base." To illustrate this situation, curves have been plotted, which show the sum of depreciation and a fair return on the respective book values. The following additional conclusions may then be derived:

9. Under entirely static conditions, the "rates" which might be claimed as fair will differ widely according to the method of depreciation. The retirement method is the most lucrative in the long run.

10. Moderate expansion tends to decrease discrepancies in "rates" considered fair under various methods. Beyond a certain limit, the relative positions of "rate" levels will be reversed. For instance if, in the case C above, expansion were to exceed the force of 6 per cent per annum,[9] the diminishing-balance method would become the most lucrative one, while the retirement method would be at the bottom in that respect.

11. Increasing replacement costs also decrease discrepancies among various fair-"rate" concepts. Figure 4 shows that, in the circumstances assumed, a combination of 5 per cent expansion and 2 per cent increase in replacement costs ultimately calls for the same "rate" under any method of depreciation. If either rate were higher, the reversal of "rate" levels would also occur.

III

The foregoing comparison of depreciation methods shows how desirable it would be to find the "true" method. The search for truth

[9] Solve for x the equation
$$-D_r(x, \infty)e^{-x\infty} + i = -D_d(x, \infty)e^{-x\infty} + iC_d(x, \infty)e^{-x\infty}.$$

starts from the economic theory of capital value. According to fundamental principles, the capital value (use value) of a single machine consists of the discounted net rental, plus the discounted scrap value. Prof. Harold Hotelling[10] formulates the concept as follows:

$$(43) \qquad V(t) = \int_{t}^{T} [zQ(\tau) - E(\tau)]e^{-\int_{t}^{\tau} i(\nu)d\nu}d\tau + s(T)e^{-\int_{t}^{T} i(\nu)d\nu}.$$

In this notation, which has been changed somewhat to avoid conflict with that adopted elsewhere in this paper, $V(t) =$ unknown capital value of a single machine, $z =$ unknown "theoretical selling price" of a unit of product, $Q(t) =$ known rate of output, $E(t) =$ known operating expenses,[11] $T =$ unknown time at which the machine ought to be discarded, $s(t) =$ known scrap value, and $i(t) =$ known rate of interest.

In accordance with the main postulate that "everything in the owner's power will be done to make $V(t)$ a maximum, when $t > 0$,"[12] T is determined by $dV(t)/dT = 0$:

$$(44) \qquad zQ(T) - E(T) = i(T)s(T) - s'(T).$$

The machine must be scrapped, when the sales exceed the operating expenses by interest on the scrap value (less the rate of change of the scrap value, if any).

A third equation is needed to find the three unknowns $V(t)$, z, and T. Prof. Hotelling obtains it by rewriting (43) for $t = 0$, "since," he says, "the value of a new machine $[V(O)]$ is its cost."[13]

This statement can no longer be accepted as generally valid. The connection between the cost and the capital value of a new machine is by no means a direct one. Its market price oscillates around a balancing point determined marginally by the least efficient pair of all producers and consumers of such machines. In the general case, therefore, capital value is apt to be greater than cost, although fluctuations may temporarily lead to the reverse. Be that as it may in a special case, it follows that, unless by pure accident the equality happens to be true, its assumption will always distort the problem. All three answers so obtained must necessarily be wrong.

The proper way to pass from the capital value $V(t)$ to the book value (unexpired cost) $B(t)$ of a single machine is to substitute the rate of

---

[10] "A General Mathematical Theory of Depreciation," *Journal of the American Statistical Association*, September, 1925.

[11] No definition of operating expenses is given, but the formula shows that, if the selling price is to be correct, the term must include every business expense except depreciation.

[12] *Op. cit.*, p. 343.

[13] *Ibid.*

profit $p(t)$ for the rate of interest. Although (43) remains the correct capital-value formula, we must use for depreciation purposes

$$(45) \qquad B(t) = \int_t^T [zQ(\tau) - E(\tau)]e^{-\int_t^\tau p(\nu)d\nu}d\tau + s(T)e^{-\int_t^T p(\nu)d\nu};$$

$$(46) \qquad B(0) = \int_0^T [zQ(\tau) - E(\tau)]e^{-\int_0^\tau p(\nu)d\nu}d\tau + s(T)e^{-\int_0^T p(\nu)d\nu};$$

$$(47) \qquad zQ(T) - E(T) = p(T)s(T) - s'(T).$$

The machine will now be replaced as soon as it ceases to earn the rate of profit on its scrap value, unless the market price of scrap varies. It is true that the capital value of a single machine may thus be kept below the maximum, but it is more important to maximize the value of the whole chain of machines succeeding each other. Only when no replacement is intended, is it correct to use (44) in lieu of (47).[14]

The simplest depreciation problem is that of a regulated monopoly, where $p(t)$ is known as the legal rate of profit. This rate can be made the actual rate by establishing the principle that discrepancies between the two create accounts receivable or payable, to be liquidated in due course. The task is then fully defined and the three equations (45) to (47) can be solved without difficulty. For public utilities at least, it appears theoretically possible to prescribe a standard method of depreciation, which will enforce the spirit of regulation. Prof. Hotelling's numerical example[15] is a fair illustration of public-utility depreciation theory as applied to a single machine, if the legal rate of profit is used in lieu of his rate of interest. It would be even more instructive, had he not omitted the scrap value for the sake of brevity. The method, however, is not applicable to competitive conditions, nor does it take into account the theory of the composite plant.

When it is attempted to apply the results of single-machine analysis to many machines installed at the same time, it becomes apparent that, if the value theory of discarding is applicable,[16] all terms under the integral sign of the book-value formula (45) must be functions of the date of scrapping T, which is given by the reversed mortality formula $T = M^{-1}(y)$. Equation (45) may be so revised and then summed up with respect to the ordinate:

[14] A variable rate of profit may be considered to pass without discontinuity from one machine to the next. This means, however, that the scrapping date of the first machine depends upon all successive replacements.

[15] *Op. cit.*, p. 347.

[16] That need not always be the case. When a machine has an almost fixed total capacity for service and can render it with almost uniform, or even increasing efficiency, discarding will have a physical, rather than a value aspect. The date of discarding should then be known from experience.

$$(48) \quad c(t) = \int_0^{M(t)} \left[\int_t^T [z(T)Q(T, \tau) - E(T, \tau)] e^{p(t-\tau)} d\tau + s(T) e^{p(t-T)} \right] dy.$$

That would be the "true" book-value formula of the original plant, to be placed alongside equations (19) to (22) for substitution in the general expression (16) of the composite book value. The composite "rate" or selling price $Z(x, q, t)$ per unit of output can also be found by the usual method of deriving composite curves. Let

$$(49) \quad \xi(t) = \int_0^{M(t)} z(T)Q(T, t)dy \quad \text{and} \quad \lambda(t) = \int_0^{M(t)} Q(T, t)dy,$$

$$(50) \quad Z(x, q, t) = \frac{\xi(t) + \int_0^t u(x, q, \tau)\xi(t - \tau)d\tau}{\lambda(t) + \int_0^t u(x, \tau)\lambda(t - \tau)d\tau}, \quad t < n.$$

This solution is oversimplified in many respects, because a change in replacement costs presumably changes the mortality curve and therefore also the renewal function. Additional variations arise, when expenses are analyzed into their component parts, each of which behaves differently, some being related to the whole enterprise, rather than to individual machines. It is impossible to consider these complications within the scope of the present article.

IV

Let us now search for the "true" method of depreciation applicable to the broader field of competitive enterprise. There, the rate of profit $p(t)$ is unknown, so that further information is needed. The first impulse is to turn to the market for an additional equation. The market price of the machine's product at any given time is determined by the rate of demand corresponding to the rate of production of the industry, subject to the cushioning effect of inventories. Cost regulates production only in the long run by influencing replacement policies. In order to recover at least a portion of the purchase price, a machine once acquired must be operated, even if the owner finds himself on the wrong side of the theoretical margin. Although the market price of the product thus depends upon the rate of output of every enterprise in a certain field, it is virtually independent of the production costs of the machines already in service. This situation apparently prevails as long as the market price is in excess of the direct expenses which can be saved by stoppage. For purposes of a general theory, the market price z is therefore best considered a known function. How to find it, is not primarily a depreciation problem, except in special cases.[17]

[17] Unregulated monopoly is not a special case in this sense, because the market

Even if the market price of the product is known, however, the task is still indefinite, because any suitable combination of book value and rate of profit will satisfy the conditions imposed. The next step which suggests itself is to introduce the changing opportunities for bargains in the market for machines. Theory carried to extremes would then postulate perfect foresight of all economic conditions up to the death of the enterprise. On such a basis, the dates of replacement and the total production surpluses of successive machines could be calculated, but the distribution of the surpluses over the lives of the respective machines, i.e., the continuous time shape of the profit, remains undefined. No matter how far analysis and conjecture are carried, it is necessary to assume the form of the profit function either deliberately or by doing—perhaps unwittingly—something equivalent. *Any depreciation method ever devised amounts merely to such an assumption.*

To support this perhaps startling conclusion, Prof. Hotelling's "general" theory will be re-examined. Its essence is evidently:

$$(51) \qquad B(t) = \int_t^T [wQ(\tau) - E(\tau)]e^{-\int_t^\tau i(v)dv}d\tau + s(T)e^{-\int_t^T i(v)dv}.$$

At first sight, this may seem to be a general way to pass from capital value to book value, because the substitution of the so-called unit cost w for the unit selling price z apparently rectifies the error committed by forcing $V(t) = B(t)$. On closer scrutiny it is found, however, that if z is considered constant for illustrative purposes, it does not follow that w must also be constant. The assumption that it is amounts to a definition of the profit function in a manner which is inconsistent with the value theory of discarding.

The proof is easily furnished by equating the truly "general" expression (45) with the special assumption (51) and solving for the profit:

$$(52) \qquad p(t) = i(t) + \frac{(z - w)Q(t)}{B(t)}.$$

When $z > w$, it follows immediately that $p(T) > i(T)$, because $B(T) > 0 < Q(T)$. The profit at the moment of discarding contradicts the theory upon which discarding is based!

---

price and the proportion of the maximal productive capacity which will yield the most remunerative market price can be found from two equations virtually independent of the other three. The first of these additional equations is evidently the demand function; the second expresses the requirement that the derivative of the net rental with respect to the unknown proportion of the maximal productive capacity must vanish.

The principle involved may be clarified further by citing a numerical example. Let $Q(t) = 100e^{-.04t}$, $E(t) = \$10e^{.20t}$, $i = .05$, $B(0) = \$200$, and $B(T) = s(T) = \$10$. These data must be substituted in Prof. Hotelling's three equations, namely (51) as written, the same equation for $t = 0$, and equation (44), in which w should be read for z. The solution is $T = 7.4771$ years, $w = \$.608392$, and

(53) $\qquad B(t) = \$675.99e^{-.04t} + \$66.67e^{.20t} - \$542.66e^{.05t}.$

If it so happens that $z = w$, all is well, because $p(t) = i$. Such an accident can not be expected, however. Let us therefore assign another value to z, say $\$.628699$. Substitution into (52) leads to the result:

(54) $\quad p(t) = .05 + \$2.0307/B(t)e^{.04t}; \; p(0) = .0603$ and $p(T) = .2006.$

The books report that the machine is earning over 20 per cent per annum on its scrap value at the moment when it is discarded for the reason that it is earning only 5 per cent! All theories of depreciation known to me fail to pass this test. Yet it is not difficult to devise a method which will behave in a consistent manner when foresight is perfect. The simplest example would be:

(55) $\;\; B(t) = \$571.54e^{-.04t} + \$76.92e^{.20t} - \$448.46e^{.07t}$ for $0 \leqq t \leqq 7.597$

and

$\qquad B(t) = \$10 \qquad\qquad\qquad\qquad$ for $7.597 \leqq t \leqq 7.6117.$

This book-value curve determines the profit function

(56) $\qquad p(t) = .07 \qquad\qquad\qquad$ for $0 \leqq t \leqq 7.597,$

and

$\qquad p(t) = 6.28699e^{-.04t} - e^{.20t} \quad$ for $7.597 \leqq t \leqq 7.6117,$

$\quad p(7.6117) = .05.$

If this had been a public-utility problem with the added condition that the machine is not to be replaced, but allowed to serve until it ceases to earn the rate of interest on its scrap value, it could be said that equation (55) is the correct solution.[18] For the competitive case,[19] it looks reasonable, because it is at least consistent, but any number of other profit curves may be subjected to the same condition $p(7.6117) = .05$. Each will produce a different book-value curve, i.e., a different depreciation theory, none of which is either the "true" or the "general"

---

[18] The calculation consists of solving simultaneously the three equations (45), (46), and (47) for $p(t) = .07$. This leads to $T = 7.597$, $z = \$.628699$ and to (55). The selling price was then substituted in equation (44) to find $T = 7.6117$. The profit function for the interval $7.597 \leqq t \leqq 7.6117$ is given by equation (47).

[19] Consider z as known and solve instead for the unknown profit $p(t) = p$.

one. All that can be demanded is that the time shape of the profit and the date of discarding be based upon a single assumption, not upon two contradictory ones. If $p_j(t)$ be the rate of profit of the jth machine in the chain of replacements and T_j its date of replacement ($j = 1, 2, 3, \cdots, \omega$), the condition $p_j(T_j) = p_{j+1}(T_j)$ must be observed until the end of operations, when it may be written $p_\omega(T_\omega) = i(T_\omega)$. Strictly speaking, $p_j{}'(T_j) = p_{j+1}{}'(T_j)$ is also desirable.

The purely theoretical value aspect of depreciation could, of course, be elaborated indefinitely. Foremost among the topics omitted are the question of obsolescence, i.e., gradual improvement in the type of machines used and the relationship between the rate of production and the productive capacity. All assumptions made are also subject to other more or less obvious limitations which cannot be enumerated here. As Walras has pointed out long ago, the solution of any economic problem ultimately resolves itself into the task of solving all other problems at the same time. In such circumstances, the last-comer is always in a strategic position. No matter how many simultaneous equations his predecessors may have set up, he can always prove their naïveté by pointing to another equation, which they ought not to have overlooked.

v

Outside the limited field of regulated monopoly, the search for the true depreciation method has apparently ended in failure, because the only clue which could be found is the rate of profit itself. This rate is necessarily indeterminate within the life periods of the machines succeeding each other. In any event, therefore, an arbitrary assumption is needed, the most highly developed form of which would be the fitting of a continuous curve to the successive blocks of production surpluses.

This situation is apt to turn the thoughts of the investigator into more practical channels. Theory is an insidious affliction, however, which is difficult to shake off. For instance, it sounds highly practical to say that no method of depreciation, haphazard, naïve, or absurd though it may be, can change the *true* value of a machine. To do that, something must be done to the machine, not to the books. The date of discarding affects the machine; depreciation is a matter of bookkeeping. Prof. Hotelling has made substantially this remark which is frequently echoed.[20]

The conclusion, unfortunately, is valid only in theory, when emphasis is placed on the word "true," which implies perfect foresight. As soon as this assumption contrary to fact is abandoned, the fallacy becomes apparent. It is best disclosed by a formula of capital value

[20] *Op. cit.*, p. 341.

based entirely on the books. By means of elementary operations, the capital-value formula (43) can easily be converted into

$$(57) \qquad V(t) = B(t) + \int_t^T [p(\tau) - i(\tau)]B(\tau)e^{-\int_t^\tau i(v)dv}d\tau.$$

Capital value equals the book value, plus the discounted excess profits.

Theoretically, this statement is true for any book value and any method of depreciation, whether based upon cost or not. In practice, it is the well-known formula for appraising the capital value of a business by past experience. The future date T, up to which the prospective purchaser is willing to look ahead, is determined largely by the rate of profit, on the ground that the more moderate it is, the longer it is likely to last. Bookkeeping accordingly has great influence upon capital value after all. The more "conservative" the book value, the lower the capital value is apt to be. That, of course, is no longer the theoretically true value, but it is the only one which has a practical significance.

Accountants immediately discard their own figures and demand an appraisal of the plant and other fixed assets, whenever they are called upon to compute capital value for the purposes of sale, reorganization, etc.[21] Apart from such occasions they adhere to their depreciation methods with the proviso that the method itself matters less than consistent adherence to it, once it has been adopted. These methods generally limit guessing to a minimum considered unavoidable in the circumstances. Guesses, so accountants feel, ought to be left to the stock market which will, in due time, accustom itself to the particular accounting methods followed by a given company.[22]

That, briefly, is the present status of depreciation theory and practice. The principal error, of which mathematical economists and accountants are equally guilty, is that both groups have so far examined the problem only in terms of a single machine, disregarding the actuarial theory of the composite plant, which introduces various new aspects. The first of these is mortality and replacement; attention to it is indispensable for the purpose of gaining a true perspective. Even more important, however, are the practical criteria of the capital value of a composite plant. In the absence of perfect foresight, appraisals by the stock market are based to a great extent upon a cursory analysis of the rate of profit in terms of its contributory causes, viz., opportunity, efficiency, risk, original book value (frequently not cost), and de-

[21] For a discussion of legal and accounting practices in determining capital value see Gabriel A. D. Preinreich, "The Law of Goodwill," *Accounting Review*, Dec., 1936 and "Goodwill in Accountancy," *Journal of Accountancy*, July, 1937.
[22] Cf. conclusions 2 and 4 reached at the end of Section II.

preciation method (frequently not all-inclusive). Apart from certain extraneous items, the value of the plant can not be distinguished from the value of an enterprise as a whole. It no longer depends upon the future services of the machines which compose the plant at any given moment, but also upon the services of those which will be installed in the future.[23] Depreciation theories must be formulated and judged in the light of the situation as it actually exists, not by assuming perfect foresight. What can be done to aid and simplify the appraisal function of the stock markets is the real problem, because market quotations are the most important manifestations of capital value.

New York, N. Y.

[23] This introduces the hitherto neglected concepts of the *expansion rate* and the *horizon* as important criteria of capital value. For a cursory discussion see Gabriel A. D. Preinreich, *The Nature of Dividends*. New York, 1935, p. 9 and appendix.

Depreciation Multiplier

THE DEPRECIATION MULTIPLIER

P. de Wolff

I Introduction

ONE of the most commonly used methods of depreciating capital goods is to write off their initial value in equal annual portions in the course of the estimated lifetime. When this lifetime is correctly estimated and when the depreciation fund is left idle its value (apart from price changes which will be left entirely out of account here) at the end of the lifetime will evidently be equal to the initial value of the capital good and will allow for the replacement of the worn-out item by an exactly identical new one.

In reality, this equality will very seldom be fulfilled. This is not only due to the fact that the individual lifetimes of identical capital goods will vary so that, at best, only a correct average can be used which could yield equality for very large values of the initial stock only, but still more because depreciation funds are usually reinvested in one way or another long before replacement has become necessary.

In particular, in dynamic processes replacement and depreciation may differ considerably. Already in 1953 E. Domar [2] has proved that in an exponentially growing economy, when depreciation funds are immediately reinvested, the method of constant depreciation over time will lead to an amount of depreciation D_t at time t bearing a constant ratio to the amount to be replaced R_t, given by:

$$D_t : R_t = (e^{\gamma T} - 1) : \gamma T \qquad (1)$$

where γ is the rate of growth of the economy and T the average lifetime of the capital goods. For reasonable values of γ and T the ratio may differ greatly from one.

Dr. Horvat [3] studies the consequences of the method for a different kind of dynamic process. He considers a large stock of identical new capital goods K_0 at $t = 0$. All items are assumed to have exactly the same lifetime, T. Hence, according to the method of constant depreciation over time, each year $\dfrac{1}{T}$ of the capital stock at the beginning of the year is depreciated. Dr. Horvat now makes two as-

sumptions: that the depreciation fund is immediately reinvested in identical new capital goods and that *no other investment takes place*. Under these circumstances, the time path of the capital stock is completely determined. At first it will be subject to heavy fluctuations but these tend to be dampened and by some experimentation with numerical data, Dr. Horvat makes it plausible that the capital stock K_t will approach a limiting value K_∞. As the ratio $M = K_\infty : K_0$ in the cases considered by him turns out to be larger than one, he introduces for M the term *depreciation multiplier*.

In this article, it will not only be proved that Dr. Horvat's conjectures are correct and that M depends on the value of n, but also that K_t approaches a limiting value for a much more general form of lifetime distribution of the capital goods. The multiplier then turns out to depend to the average, T, and the standard deviation, σ, of this distribution and moreover, for certain combinations of the two, M can even become less than one.

The techniques to be used are the familiar ones of renewal theory. They go back to work done by Lotka in 1939 [4]. Here, the discrete approach will be used. Reference may also be made to a book by Cox [1] which contains bibliographical material.

II The General Formula for the Depreciation Multiplier

The capital stock (measured in units) at the end of time unit t will be indicated by K_t, the amount of depreciation (equal to reinvestment) at the same point of time by R_t. The simplifying assumption will be made that the lifetime of all the individual capital goods is a whole number and that at the end of time unit i a fraction, a_i, of the initial stock wears out. The maximum lifetime will be indicated by n, so that:

$$\sum_1^n a_i = 1. \qquad (2)$$

From this it follows that the average lifetime, T, of the capital goods is given by:

$$T = \sum_{1}^{n} i a_i \tag{3}$$

At an arbitrarily chosen point of time t (t is a whole number) for which $t > n$ the amount of capital in use consists of what is left from the reinvestments at t, $(t - 1)$, ... $(t - n + 1)$. Hence:

$$K_t = R_t + (1 - a_1)R_{t-1} + \\ (1 - a_1 - a_2)R_{t-2} + \cdots + a_n R_{t-n+1}. \tag{4}$$

On the other hand, due to the depreciation method adopted, R_{t+1} is equal to:

$$R_{t+1} = \frac{1}{T} K_t. \tag{5}$$

From (4) and (5) it follows that R_t satisfies the following linear difference equation with constant coefficients:

$$TR_{t+1} = R_t + (1 - a_1)R_{t-1} + \\ (1 - a_1 - a_2)R_{t-2} + \cdots + a_n R_{t-n+1}. \tag{6}$$

Introducing the notation:

$$D_i = \sum_{i}^{n} a_j, (i = 1, 2, \cdots, n) \tag{7}$$

(with $D_1 = 1$ and $D_i \geqq D_{i+1}$)

(6) can be written as:

$$TR_{t+1} = D_1 R_t + D_2 R_{t-1} + \cdots + D_n R_{t-n+1}. \tag{8}$$

The characteristic equation of this linear difference equation reads as follows:

$$T\lambda^n = D_1\lambda^{n-1} + D_2\lambda^{n-2} + \cdots + D_n. \tag{9}$$

As

$$\sum_{1}^{n} D_i = \sum_{1}^{n} \sum_{i}^{n} a_j = \sum_{1}^{n} i a_i = T, \tag{10}$$

it is immediately clear that (9) has one root equal to one. Moreover, it can easily be proved that (9) can have no roots λ with modulus > 1 or with modulus $= 1$ but with $\lambda \neq 1$. To this end, suppose that λ_j, with $|\lambda_j| = \rho \geqq 1$ and $\lambda_j \neq 1$, would satisfy (9).

It then follows from (9) that:

$$T|\lambda_j|^n = T\rho^n = |D_1\lambda_j^{n-1} + D_2\lambda_j^{n-2} \\ + \cdots + D_n|.$$

Further:

$$|D_1\lambda_j^{n-1} + D_2\lambda_j^{n-2} + \cdots + D^n| \leqq D_1\rho^{n-1} \\ + D_2\rho^{n-2} + \cdots + D_n \leqq \rho^{n-1} \sum_{1}^{n} D_i.$$

In this relation, the first equality sign can only hold when arg. $\lambda_j = 0$, the second when $\rho = 1$. Both signs can therefore hold only when $\lambda_j = 1$. This case, however, is excluded. Hence, in view of (10), $T\rho^n < T\rho^{n-1}$ or $\rho < 1$ which clearly contradicts the assumptions made. Now if we denote the roots of (9) by 1, λ_1, λ_2, ..., λ_{n-1}, the general solution of (8) will be of the following form:

$$R_t = c_0 + c_1\lambda_1^t + c_2\lambda_2^t + \cdots + c_{n-1}\lambda_{n-1}^t. \tag{11}$$

Here c_0, ..., c_{n-1} are constants depending on the initial conditions of the problem. As the moduli of all λ_i are < 1 the terms $c_i\lambda_i^t$ will vanish when $t \to \infty$. Hence

$$R_t \to c_0 \text{ if } t \to \infty. \tag{12}$$

This proves that under all circumstances, even for the general form of the lifetime distribution considered here, R_t approaches a limiting value and, consequently, that a multiplier does exist.[1] Its value will now be computed.

R_t is completely determined when in addition to the λ_i the n constants, c_i, are given. They can be computed when R_t is given for n different values of t. Obviously, (4) and (5) are satisfied by putting $R_0 = K_0$ and $R_t = 0$ for $t = -1, -2, \ldots, (-n + 1)$. Substituting these values in (11) the following set of equations is obtained:

[1] The speed at which R_t (and consequently K_t) approaches its limit obviously is dependant on the size of the moduli of the roots λ_i. The smaller these values, the more rapidly will the terms $c_i\lambda_i^t$ die away as t increases. It is not easy to give an effective general formula for the upper bound of the λ_i. It is immediately clear that all $\lambda_i = 0$ when $D_i = 0$ for $i > 1$ (as the lifetime distribution is concentrated at the time point one).

On the other hand it is clear that R_t is subject to very heavy fluctuations when all $D_1, D_2, \ldots, D_n$ are equal (i.e., as the centre of gravity of the distribution has entirely shifted to the time point n). One gets the impression that the moduli of λ_i tend to increase when the centre shifts from 1 to n. Even in that particular case it is not easy to find a general formula for the upper bound. For $n = 4$ and 10 the maximum value of $|\lambda_i|^n = 0.116$ and 0.13g showing that in these particular cases even the contribution of the larger roots decreases considerably in the course of the maximal life span n.

$$c_0 + c_1 \qquad + c_2 \qquad + \cdots + c_{n-1} \qquad = K_0$$
$$c_0 + c_1\lambda_1^{-1} \qquad + c_2\lambda_2^{-1} \qquad + \cdots + c_{n-1}\lambda_{n-1}^{-1} = 0 \qquad (13)$$
$$c_0 + c_1\lambda_1^{-(n-1)} + c_2\lambda_2^{-(n-1)} + \cdots + c_{n-1}\lambda_{n-1}^{-(n-1)} = 0.$$

From this set c_0 can be derived by the rule of Cramer as:

$$c_0 = K_0 \begin{vmatrix} \lambda_1^{-1} \cdots\cdots \lambda_{n-1}^{-1} \\ \\ \lambda_1^{-(n-1)} \cdots \lambda_{n-1}^{-(n-1)} \end{vmatrix} : \begin{vmatrix} 1 & 1 & \cdots & 1 \\ 1 & \lambda_1^{-1} & \cdots & \lambda_{n-1}^{-1} \\ 1 & \lambda_1^{-(n-1)} \cdots & \lambda_{n-1}^{-(n-1)} \end{vmatrix}. \qquad (14)$$

Putting $\lambda_j^{-1} = \mu_j$, $(j = 1, 2, \ldots, n-1)$ it is clear that the numerator of (14) is divisible by $\mu_1\mu_2 \ldots \mu_{n-1}$. After removing this factor both the numerator and the denominator are determinants of the so-called Vandermonde type expandable as continuous products of all the pair-wise differences of the μ_i (respectively, μ_i and 1 for the denominator). Consequently (14) can be written as follows:

$$c_0 = K_0 \frac{\mu_1 \cdots \mu_{n-1}(\mu_{n-1} - \mu_1)(\mu_{n-1} - \mu_2) \cdots (\mu_{n-1} - \mu_{n-2})(\mu_{n-2} - \mu_1) \cdots (\mu_2 - \mu_1)}{(\mu_{n-1} - 1)(\mu_{n-1} - \mu_1) \cdots (\mu_{n-1} - \mu_{n-2})(\mu_{n-2} - 1) \cdots (\mu_1 - 1)}$$
$$= K_0 \frac{\mu_1 \cdots \mu_{n-1}}{(\mu_{n-1} - 1) \cdots (\mu_1 - 1)} = \frac{K_0}{(1 - \lambda_1)(1 - \lambda_2) \cdots (1 - \lambda_{n-1})}. \qquad (15)$$

Now the λ_j satisfy the equation resulting from (9) by eliminating the root one. Taking into account (10)

$$T\lambda^n - D_1\lambda^{n-1} - \cdots - D_n = D_1(\lambda^n - \lambda^{n-1})$$
$$+ \cdots + D_n(\lambda^n - 1) = (\lambda - 1)[D_1\lambda^{n-1}$$
$$+ \cdots + D_n(\lambda^{n-1} + \lambda^{n-2} + \cdots + 1)]$$
$$= (\lambda - 1)[(D_1 + D_2 + \cdots + D_n)\lambda^{n-1}$$
$$+ (D_2 + \cdots + D_n)\lambda^{n-2} + \cdots + D_n].$$

Hence:

$$(D_1 + D_2 + \cdots + D_n)\lambda^{n-1} + (D_2 + \cdots + D_n)\lambda^{n-2} + \cdots + D_n = T(\lambda - \lambda_1)$$
$$(\lambda - \lambda_2) \cdots (\lambda - \lambda_{n-1}). \qquad (16)$$

Putting $\lambda = 1$, (16) reduces to:

$$(D_1 + D_2 + \cdots + D_n) + (D_2 + \cdots + D_n)$$
$$+ \cdots + D_n = T(1 - \lambda_1)(1 - \lambda_2)$$
$$\cdots (1 - \lambda_{n-1}).$$

From this result it follows, that:

$$c_0 = \frac{TK_0}{D_1 + 2D_2 + \cdots + nD_n}. \qquad (17)$$

According to (5) $K_t = TR_{t+1}$, hence
$$K_t \rightarrow Tc_0 \text{ as } t \rightarrow \infty$$
and, finally,

$$M = \frac{K_\infty}{K_0} = \frac{T^2}{D_1 + 2D_2 + \cdots + nD_n}. \qquad (18)$$

In view of (7):

$$D_1 + 2D_2 + \cdots + nD_n = (a_1 + a_2 + \cdots + a_n) + 2(a_2 + \cdots + a_n) + \cdots + na_n$$
$$= \frac{1}{2} \sum_1^n a_i \, i(i + 1). \qquad (19)$$

This leads to the required result:

$$M = \frac{2T^2}{\sum_1^n a_i \, i(i + 1)}. \qquad (20)$$

For various applications it is convenient to express the denominator of (20) as a function of T and σ^2, the variance of the lifetime distribution.

$$\sum_1^n a_i \, i(i + 1) = \sum_1^n a_i \, (i - T)^2 + 2T \sum_1^n a_i \, i$$
$$- T^2 \sum_1^n a_i + \sum_1^n a_i \, i.$$

Applying (2) and (3) and observing that

$$\sigma^2 = \sum_1^n a_i(i - T)^2 \qquad (21)$$

(20) reduces to:

$$M = \frac{2T^2}{\sigma^2 + T(T + 1)}. \qquad (22)$$

III Some Special Cases

It is interesting to apply the formulas (20) and (22) to a number of special cases.

i) *All Individual Items Have the Same Lifetime n*

In this case,
$$a_1 = a_2 = \cdots = a_{n-1} = 0, a_n = 1.$$
It follows immediately that $T = n$ and $\sigma = 0$ and (22) reduces to

$$M = \frac{2n}{n+1}. \tag{23}$$

This is the case considered by Dr. Horvat. For $n = 3, 4, 5$, and ∞, (23) leads to 1.5, 1.6, 1.67, and 2 in excellent agreement with his empirically derived results. The result for $n = \infty$ is only interesting because for moderate values of n, M comes very close to 2. The limit $n = \infty$ itself, however, lacks significance. In general, K_t will only come close to MK_0 for values of $t >> n$ which evidently is impossible when n itself goes to infinity. In effect, when $T \to \infty$ depreciation goes to zero and the capital stock will permanently remain equal to its initial value (and not approach the value $2K_0$).

As a matter of fact, Dr. Horvat deduced his results from a difference equation expressed in K_t and not in R_t. This equation can easily be derived as follows. In his case (4) reduces to:

$$K_t = R_t + R_{t-1} + \cdots + R_{t-n+1}. \tag{24}$$

Hence,

$$K_{t+1} - K_t = R_{t+1} - R_{t-n+1}.$$

Applying (5) this result can (as $T = n$) be transformed into:

$$nK_{t+1} - nK_t = K_t - K_{t-n}$$

or

$$nK_{t+1} = (n+1)K_t - K_{t-n} \tag{25}$$

This is Dr. Horvat's difference equation. Its characteristic equation is equal to:

$$n\lambda^{n+1} = (n+1)\lambda^n - 1. \tag{26}$$

In addition to the $(n-1)$ roots λ_j of

$$n\lambda^n = \lambda^{n-1} + \cdots + 1 \tag{27}$$

which is the form taken by (9) in this case, (27) has a double root equal to 1. This leads to a term $C_0 + C_0^* + t$ in the general solution. This result, however, is irrelevant to the problem considered here, as the second root, 1, clearly has been introduced by the transition of R_t to K_t.

ii) *The Lifetime Distribution is Rectangular*

In this case $a_1 = a_2 = \cdots = a_n = \dfrac{1}{n}$.

It follows that:

$$T = \sum_1^n ia_i = \frac{1}{n} \cdot \frac{1}{2} n (n+1) = \frac{1}{2}(n+1)$$

$$\sigma^2 = \frac{1}{n}\sum_1^n i(i+1) = \frac{1}{3}(n+1)(n+2).$$

Hence:

$$M = \frac{2 \cdot [\frac{1}{2}(n+1)]^2}{\frac{1}{3}(n+1)(n+2)} = \frac{3}{2} \cdot \frac{n+1}{n+2}. \tag{28}$$

It is easily proved that in this case M is smaller than the M of case (i) for every value of n.

For $n \to \infty \qquad M \to \dfrac{3}{2}$.

iii) *The Lifetime Distribution Has Only Two Nonvanishing Frequencies*

It will be assumed that only a_1 and $a_n \neq 0$.

Putting $a_1 = a$ it follows from (2) that $a_n = 1 - a$.

In this case:

$$T = a + (1-a)n = n - (n-1)a \tag{29}$$

$$\sum_1^n a_i i(1+i) = 2a + (1-a)n(n+1)$$
$$= n(n+1) - (n-1)(n+2)a. \tag{30}$$

Hence:

$$M = \frac{2[n - (n-1)a]^2}{n(n+1) - (n-1)(n+2)a}. \tag{31}$$

This case is interesting for various reasons. In the first place, it can be considered as a modification of (i). All individual items have lifetimes equal to n apart from a fraction of "early failures" with lifetime 1. When a is small compared to 1, as is to be expected in a case of early failures, (31) can be expanded as a power series of a. Neglecting all powers higher than the first, (31) yields:

$$M = \frac{2n}{n+1}\left[1 - \frac{n-1}{n+1} \cdot a\right]. \tag{32}$$

A second more important reason is that for given values of n and T the variance σ^2 obtains its maximum value when only the extreme frequencies a_1 and $a_n \neq 0$. This result will be used in section IV.

iv) *The Lifetime Distribution is a Geometric Series*

It has already been explained that the multiplier has a meaning only when n is finite. Hence, in this case the series has to be finite also. The corresponding formulae, however, become rather clumsy and it is easy to see that the infinite case still has some importance. In this case, the fraction of the capital stock wearing out each year is equal to $\frac{1}{T}$ times this stock and consequently equal to the depreciation The capital stock therefore remains constant over time and obviously $M = 1$.

This result can be used as a check of formula (20). Representing the ratio of the series by τ, a_i must be equal to

$$a_i = c \cdot \tau^i \qquad (i = 1, 2, \cdots \text{ ad infinitum}).$$

The constant c has to be chosen in such a way that $\overset{0}{\underset{1}{\Sigma}} a_i = 1$, hence $c = \frac{1 - \tau}{\tau}$.

In this case:

$$T = \overset{0}{\underset{1}{\Sigma}} \frac{1 - \tau}{\tau} \cdot i\tau^i = \frac{1}{1 - \tau}$$

$$\sigma^2 = \overset{0}{\underset{1}{\Sigma}} \frac{1 - \tau}{\tau} i(i + 1)\tau^i = \frac{2}{(1 - \tau)^2}.$$

Hence:

$$M = \frac{2 \dfrac{1}{(1 - \tau)^2}}{2 \dfrac{1}{(1 - \tau)^2}} = 1, \qquad (33)$$

as was expected.

IV The Extreme Values of M

In the preceding section, M has been computed for a number of special cases. These results, however, do not yet show the extreme values of M nor the conditions under which they are reached. This problem can be solved as follows. It has already been mentioned in section III (iii) that for given n and T the variance σ^2 is maximal when only a_1 and $a_n \neq 0$. Calling these frequencies again a and $1 - a$, it follows from (29) that

$$a = \frac{n - T}{n - 1} \text{ and } 1 - a = \frac{T - 1}{n - 1}; \qquad (34)$$

σ^2 then becomes

$$\frac{n - T}{n - 1} (T - 1)^2 + \frac{T - 1}{n - 1} (n - T)^2$$

$$= (n - T)(T - 1). \qquad (35)$$

Hence for given n and T:

$$0 \leq \sigma^2 \leq (n - T)(T - 1). \qquad (36)$$

On the other hand it follows from (22) that:

$$\sigma^2 = - T + \frac{2 - M}{M} T^2. \qquad (37)$$

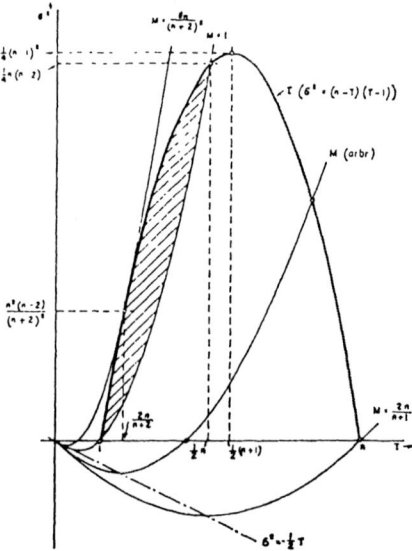

FIGURE 1 — THE RELATIONS BETWEEN M, T, and σ

The two relations (36) and (37) determining all possible combinations of M, T, and σ^2 must be represented by points inside or on the boundary of the area delimited by the parabola I and the T-axis. The combinations of T and σ leading to the same value of M are indicated by (37). Each value of M produces a parabola with a vertical axis which passes through the origin and has its summit on the line $\sigma^2 = -\frac{1}{2} T$. It is immediately clear that the maximum of M will be obtained when the parabola from the pencil represented by (37) passes through the point $(T = n, \sigma^2 = 0)$, leading to $M = \frac{2n}{n + 1}$. This shows that the case considered by Dr. Horvat leads to the maximal value of the multiplier M. On the other hand, when $\sigma^2 = 0$ and T reaches its lowest value, as 1, M too is equal to 1, but this is not the lowest possible value of M. The value $M = 1$ will in fact be reached for a whole series of combinations of T and σ^2 reaching from $(T = 1, \sigma^2 = 0)$ at one end to $\left[T = \frac{1}{2} n, \right.$

$\sigma^2 = \dfrac{n(n-2)}{4}$] at the other end. The minimum value of M is obtained when the parabola (37) just touches the one representing (36). This requires that the equation:

$$- T + \frac{2 - M}{M} T^2 = (n - T)(T - 1)$$

or

$$\frac{2}{M} T^2 - (n + 2)T + n = 0 \qquad (38)$$

has equal roots.
Therefore its discriminant is:

$$(n + 2)^2 - 4n \cdot \frac{2}{M} = 0$$

or

$$M = \frac{8n}{(n + 2)^2}. \qquad (39)$$

This value goes to zero as $n \to \infty$ and already for a moderate value of $n = 20$ it is equal to $\sim \dfrac{1}{3}$.

In this case,

$$\sigma^2 = \frac{n^2(n - 2)}{(n + 2)^2} \qquad (40)$$

$$T = \frac{2n}{n + 2} \qquad (41)$$

$$a = \frac{n^2}{(n - 1)(n + 2)}. \qquad (42)$$

For large values of n the minimum M will be reached for $T \sim 2$, whereas $a \sim 1$, a case exactly opposite to that considered in III (iii). It should be noted, however, that $M < 1$ everywhere inside the shaded area of the parabola I and there lifetime distributions can be found which are less improbable than the one leading to the lowest possible value of M.

V Generalizations

It is possible to generalize the results of the preceding sections in various ways. Two of such possibilities will be briefly mentioned here, namely, the case that the stock consists of several varieties of capital goods each with its own characteristic lifetime distribution and the effect of an error in the estimate of the average lifetime upon which the use of the depreciation is based.

i) *Various Categories of Capital Goods*

This case can be very easily dealt with. Assuming that each category, s, is depreciated correctly according to its own average lifetime, T_s, the linearity of equation (8) guarantees that each category leads its life independent of the other.

Therefore, for each category there is a multiplier M_s so that:

$$\frac{K^s_\infty}{K^s_0} = M_s. \qquad (43)$$

The total stock at time $t = 0$ is given by

$$K_0 = \Sigma K^s_0.$$

A similar relation holds for the limiting values:

$$K_\infty = \Sigma K^s_\infty.$$

Hence:

$$M = \frac{K_\infty}{K_0} = \frac{\Sigma K^s_\infty}{\Sigma K^s_0} = \frac{\Sigma K^s_0 M_s}{\Sigma K^s_0}. \qquad (44)$$

The total multiplier M is simply equal to the weighted average of the individual ones where the initial values of the stock of each category act as weights.

ii) *The Effect of Incorrect Estimates of T*

If depreciation is computed according to an estimated lifetime T^* differing from the actual one T, the development of R_t will be described by a linear difference equation which can be derived from (8) by replacing T by T^*.

This equation leads to a characteristic equation of the following form:

$$T^*\lambda^n = D_1\lambda^{n-1} + D_2\lambda^{n-2} + \cdots + D_n. \qquad (45)$$

This equation does not have a root equal to 1. However, it still has a real positive root λ_0 and it is easy to see that this root is < 1 when $T^* > T$ and > 1 when $T^* < T$. In both cases, this root will be bigger than the modulus of any of the other roots. This can be proved by a slight modification of the proof given in II. Suppose that λ_j, with $|\lambda_j| = \rho > \lambda_0$, would satisfy (45). Then:

$$T^*\lambda_j = D_1 + D_2\lambda_j^{-1} + \cdots + D_n\lambda_j^{-(n-1)}.$$

Hence:

$$T^*_\rho = | D_1 + D_2\lambda_j^{-1} + \cdots + D_n\lambda_j^{-(n-1)} |$$
$$< D_1 + D_2\rho^{-1} + \cdots + D_n\rho^{-(n-1)} < D_1$$
$$< D_1 + D_2 + \cdots + D_n \qquad < D_1$$
$$+ D_2\lambda_0^{-1} + \cdots + D_n\lambda_0^{-(n-1)} < T^*\lambda.$$

This is clearly a contradiction.

Therefore also when $T^* \neq T$ the real root determines the movement of R_t and K_t in the long run. However, there does no longer exist a multiplier. If $T^* < T$, $K_t \to \infty$; if $T^* > T$, $K_t \to 0$. But in both cases, λ_0 finally determines the pattern of development.

Obviously, it is not possible to give a general formula for λ_0 but when $T^* - T = \Delta T$ is small compared to T it is possible to derive an approximate expression for λ_0. Differentiating both sides of (45) with respect to T^*, considering λ as a function of T, the following result is obtained:

$$\lambda^n \Delta T + n\lambda^{n-1} \cdot T\Delta\lambda = [D_1(n-1)\lambda^{n-2} + \cdots + D_{n-1}]\Delta\lambda.$$

For $T^* = T$ and $\lambda = 1$ this leads to:

$$- \Delta T = [nT - (n-1)D_1 - \cdots D_{n-1}]\Delta\lambda$$
$$= [D_1 + 2D_2 + \cdots + nD_n]\Delta\lambda$$

or, in view of (19),

$$\Delta\lambda = -2\frac{\Delta T}{\Sigma a_i i(i+1)}. \qquad (46)$$

As $(1 + \Delta\lambda)^t \sim e^{\Delta\lambda \cdot t}$ the assymptotic growth rate γ becomes approximately equal to:

$$= -2\frac{\Delta T}{\Sigma a_i i(i+1)} = -2 \cdot \frac{\Delta T}{T} \cdot \frac{\Sigma a_i i}{\Sigma a_i i(i+1)}. \qquad (47)$$

By an argument similar to that used in section IV it can easily be proved that for given T the factor $\dfrac{\Sigma a_i i}{\Sigma a_i i(i+1)}$ lies between $\dfrac{1}{T+1}$ and $\dfrac{T}{(n+2)T - n}$. Both limits coincide for $T = 1$ and $T = n$. Hence the effect of a relative error in T of a given size is smallest when $T = n$. Then (47) reduces to:

$$\gamma = -2 \cdot \frac{\Delta T}{n+1}. \qquad (48)$$

REFERENCES

[1] D. R. Cox, *Renewal Theory* (London, 1961).
[2] E. D. Domar, "Depreciation, Replacement and Growth," *Economic Journal*, 63 (March 1953), 1–33.
[3] B. Horvat, *Towards a Theory of Planned Economy* (Yugoslav Institute of Economic Research, Belgrade 1964), 140–148; cf. also: B. Horvat, The Depreciation Multiplier and a Generalised Theory of Fixed Capital Costs, *The Manchester School* (1958), 136–159.
[4] A. J. Lotka, "A Contribution to the Theory of Self-Renewing Aggregates, with Special Reference to Industrial Replacement," *Annals of Mathematical Statistics*, X (March 1939), 1–25.

MANAGEMENT SCIENCE
Vol. 13, No. 5, January, 1967
Printed in U.S.A.

ON THE CONVERGENCE OF PERIODIC REINVESTMENTS BY AN AMOUNT EQUAL TO DEPRECIATION[*][1]

YUJI IJIRI[2]

Stanford University

This paper analyzes the pattern of periodic reinvestments in depreciable assets when the amount of reinvestment in each period is set equal to the amount of depreciation for the period and derives a constant to which the periodic investments converge. The expression for the constant which depends upon the depreciation method is then interpreted in terms of the firm's investment and financing policy as well as its growth rate.

I. Introduction

1. Periodic Reinvestment Based on Depreciation

Depreciation is simply a method of allocating the cost of fixed assets over the useful lives of these assets and has no direct implications to the actual flow of funds in the firm. However, since depreciation is treated as a deduction in arriving at net income, it directly affects the firm's income figure which, in turn, influences the actual funds flow of the firm by means of such factors as income taxes, dividend policy, negotiation with a labor union, etc. It is, therefore, a near-sighted argument to say that depreciation has nothing to do with the funds flow of the firm.

Theoretically, if we follow the Hicksian concept of income—see [7]—we would define income of the firm to be the maximum amount that the firm can distribute during a period and still expect to be as well off at the end of the period as the firm was at the beginning. Therefore, if the firm distributes at the end of a period its entire income during the period, we would expect that the firm will be exactly as well off at the end of the period as it was at the beginning. We would expect this to be true especially if the ending balance sheet after distributing the entire income during the period is identical, account by account, with the beginning balance sheet. In particular, if we focus our attention on depreciable assets,[3] assuming the levels of all other assets and liabilities are held constant, we would expect that the same amount of the book-value of depreciable assets at the end

---

[*] Received September 1965 and revised July 1966.

[1] This study was supported, in part, by funds made available by the Ford Foundation to the Graduate School of Business, Stanford University. However, the conclusions, opinions and other statements in this publication are those of the author and are not necessarily those of the Ford Foundation.

[2] The author is indebted to Professors N. C. Churchill, W. W. Cooper, C. T. Horngren, R. K. Jaedicke, and A. H. Meltzer for their helpful comments and suggestions; in particular, to Professor Cooper for his detailed review of the manuscript which resulted in its marked improvement.

[3] See [10] for the distinction between depreciable assets and non-depreciable assets.

321

of the period as at the beginning means the same "well-offness" of the firm at the end as at the beginning.

Of course, in practice, this is not necessarily true. We must take account of the asset composition within an account as well as prices of assets. However, in the case of depreciable assets there is another important factor, depreciation, which influences the "well-offness" of the firm under the assumption of a constant book-value of depreciable assets. To maintain the constant book-value of depreciable assets means that the firm must invest in depreciable assets during a period by an amount exactly equal to depreciation plus the book-value of all depreciable assets disposed during the period. If there is no disposal of assets, then depreciation is the only factor which determines the amount of reinvestment in depreciable assets under the assumption of a constant book-value of depreciable assets.

What, then, is the relationship between the depreciation method and the pattern of reinvestment under the assumption of the constant book-value of depreciable assets? Can a firm increase its production capacity without increasing the book-value of the assets and without changes in the prices of the assets?

To state this more precisely, suppose that a firm adopts the following investment policy:

   (i) Invest in fully depreciable assets at the end of Period 0 by a given amount.

  (ii) Reinvest in the fully depreciable assets at the end of each subsequent period by an amount that is exactly equal to depreciation for the period.

 (iii) Apply the same depreciation method with a constant life to all fully depreciable assets that are acquired by the firm.[4]

For convenience we shall hereafter refer to this as a PRD (Periodic Reinvestment based on Depreciation) policy. Under this PRD policy with a given depreciation method, what will the pattern of periodic investments be in the long run?

2. PRD Convergence

For example, suppose that a firm invests $1 (it may be $1,000, $100,000, or $1,000,000, etc.) at the end of Period 0 in fully depreciable assets with a two year life (throughout the paper a year and a period are used interchangeably). Suppose further that this firm adopts the straight line depreciation method, and thus depreciates $.5 in Period 1 and $.5 in Period 2. At the end of Period 1, then, the firm can reinvest $.5 without worsening the position of non-depreciable assets even if the entire net income in Period 1 was distributed.[5] Thus, suppose that the firm actually reinvests this $.5 in fully depreciable assets which also have a two year life. Depreciation in Period 2 is then found to be $.75 ($.5 for the investment in Period 0 and $.25 for the investment in Period 1). Again the firm reinvests in fully depreciable assets with a two year life by an amount equal to $.75 at the end of Period 2, and so on. As shown in the table below, the amplitude of these periodic reinvestments decreases in the later years and the amount of reinvestment eventually converges to a constant.

3. The PRD Constant

Laya [14] showed through a numerical simulation that this constant to which periodic reinvestments converges is $.6666 \cdots$ in this example. He also prepared, based on a computer simulation, a table of constants to which periodic reinvestments converge under straight line, sum-of-years-digit, and declining balance depreciation methods for various lengths of asset lives. This table of constants suggests that a more analytical approach to the problem might be possible.

4. The PRD Formula

Indeed, it turns out that there is a very simple formula for this constant (expressed as a ratio to the initial investment) to which periodic reinvestments will converge under a PRD policy. The formula for this constant which we shall call the *PRD constant* will be described in this section and some of its implications for a firm's investment policy will also be examined. After this has been done in the present section, the remaining two sections will be devoted to a mathematical derivation of the formula.

Let us denote by d_i, $i = 1, 2, \cdots, n$, the proportion of an investment to be depreciated i periods after the investment has been made. In the example in Table 1 $d_1 = .5$ and $d_2 = .5$ since 50% each of the original investment is to be depreciated in each of the 2 periods after the investment has taken place. In the case of 10 year straight line depreciation, $d_1 = d_2 = \cdots = d_{10} = .1$. Since assets are to be fully depreciated, by assumption, it is required that the sum of all d_i's be equal to 1 and, by the nature of depreciation, none of the d_i's can be negative. By virtue of these considerations, the PRD constant, a, is given by:

$$(1) \qquad a = 1/(d_1 + 2d_2 + 3d_3 + \cdots + nd_n) = 1/\sum_{i=1}^{n} id_i .$$

For example, in the case shown in Table 1 we have

$$a = 1/(d_1 + 2d_2) = 1/(.5 + 2 \times .5) = .666 \cdots$$

which agrees with Laya's result [14]. Similarly, under a 10 year straight line depreciation method, the PRD constant is given by

$$a = 1/(.1 + 2 \times .1 + 3 \times .1 + \cdots + 10 \times .1) = .18.$$

That is, if (a) the firm invests $100,000 initially, and (b) follows the PRD policy, using the 10 year straight line depreciation method, then (c) it will ultimately be investing approximately $18,000 in each period in the long run. To compare this result with the one secured from a 10 year sum-of-years-digits method, we substitute $d_1 = 10/55$, $d_2 = 9/55$, $d_3 = 8/55$, $\cdots$, $d_{10} = 1/55$ in (1), obtaining $a = .25$. Namely, the firm with the 10 year sum-of-years-digits method and with the initial investment of $100,000 will be investing approximately $25,000 in

TABLE 1

The Pattern of Depreciation and Reinvestment

Investment in Period	0	1	2	Depreciation in Period 3	4	5	6
0	(1)	.5	.5				
1		(.5)	.25	.25			
2			(.75)	.375	.375		
3				(.625)	.3125	.3125	
4					(.6875)	.34375	.34375
5						(.65625)	.328125
6							(.671875)

each period in the long run under the PRD policy. Evidently each of these methods has quite different financial implications.

Notice that, in addition to its simplicity, an interesting aspect of formula (1) is that it is applicable to any depreciation method. This is true since all such methods can be translated into a set of proportions to be depreciated in each period once the life of the assets is fixed. We shall elaborate on this in more detail in the next section by means of "depreciation vectors."

5. Implications

What is the significance of formula (1) for the PRD constant in terms of the firm's investment and depreciation policy? Let us consider the following example. A firm initially purchases 100 machines at $1,000 each. These machines have a 10 year life with no salvage value at the end of the 10 year period. Assume that the firm reinvests in the same machines at the end of each year by an amount equal to depreciation for the year calculated on the straight line method. From the above analysis, we know that the amount of annual investments is approximately equal to $18,000 in the long run. This means that if the price of the machine remains at $1,000 each, the firm will purchase 18 machines a year. This implies that the firm will be operating with 180 machines in the long run.[6]

This information on the future capacity of the firm can be very significant and

---

[6] This increase in the machine capacity is obtained at the expense of a decrease in the machine capacity in the future. To show this by a simple example, consider the case in Table 1 where the machine life is two years and the straight line method is adopted. If the firm starts with 100 machines, the firm has a machine capacity of 100 for both Period 1 and Period 2. At the end of Period 1 the firm purchases 50 machines under the PRD policy. Then the machine capacity for Period 2 is increased to 150 but the same for Period 3 is decreased to 50. It can easily be shown that the simple sum of the machine capacity for the current and the future periods is always a constant, i.e., the initial machine capacity times the life of the machine. Thus, the PRD policy results in a shift in the machine capacity from later periods to earlier periods. In general, the firm would be better off as a result of such a shift, considering the time value of services.

valuable in analyzing the firm's investment and financing policy and growth rate. If the capacity represented by 180 machines (18 machines in each age category) is not satisfactory, then either an additional fund must be provided by reserving a part of net income or by issuing stock, or the firm must resort to various means of external financing. Similarly, if the firm expects an increase in the price of the machine beyond $1,800 each, then we know that it cannot afford to distribute the entire net income insofar as it wants to maintain the original capacity of 100 machines without the help of other financing arrangements.

Furthermore, based on the formula (1), it is clear that under the PRD policy with the same length of asset lives a firm with the declining balance method grows more than a firm with the sum-of-years-digits method which, in turn, grows more than a firm with the straight line method.

Thus, the information derived from formula (1) should be useful for many types of accounting analyses that bear upon depreciation, flow-of-funds, and investment analyses, as well as for arriving at a better understanding of the literature of depreciation.[7] Moreover, the availability of a simple formula for these processes can also be useful in considering the issues of national economic policy.[8]

II. Derivation of the PRD Formula

1. Depreciation Vector

The remainder of this paper is devoted to the derivation of formula (1). For this purpose we start by first defining a depreciation method in a more general but precise manner by introducing a concept that we shall refer to as a *depreciation vector* d in the form of row components d_i $(i = 1, 2, \cdots, n)$, i.e.,

$$(2) \qquad d \equiv (d_1, d_2, \cdots, d_n).$$

Here, as mentioned in the previous section, d_i represents the proportion of the investment that is to be depreciated i periods after the investment was made, satisfying

$$(3) \qquad \sum_{i=1}^{n} d_i = 1,$$

$$(4) \qquad d_i \geqq 0 \qquad (i = 1, 2, \cdots, n).$$

Notice that any depreciation method can be translated into a depreciation vector once the asset lives have been determined. For example, a straight line method over $n = 5$ periods is translated into the depreciation vector $(.2, .2, .2, .2, .2)$; a sum-of-years-digit method over $n = 4$ periods is represented by $(.4, .3, .2, .1)$, etc. A declining balance method over $n = 3$ periods with 10% salvage value produces depreciation equal to .54, .25, and .11 of the acquisition cost in each period, or .60, .28, and .12 of the depreciable portion (90%) of the acquisition cost. Hence, the depreciation vector is $(.60, .28, .12)$ to be applied to the depreciable

---

[7] See, e.g., [5, Ch. 14].

[8] See, e.g., [3] and [11]. This topic is also related to other well known issues on internal versus external financing—as was called to the attention of the author by A. H. Meltzer. See, e.g., Sametz [15] and the later comments by Shapiro and White in [16].

portion of the assets. Furthermore, the following analysis is not restricted only to such customary methods and is applicable even for depreciation vectors such as $(.3, 0, 0, .5, .2)$, $(0, 0, 0, 1,)$ $(1, 0, 0, 0, 0)$, etc.

Before proceeding further, however, it is necessary to resolve an ambiguity in our vector representation of depreciation methods. Consider two depreciation vectors such as $(.5, .5)$ and $(.5, .5, 0, 0, 0)$. Obviously, both vectors produce an identical stream of periodic reinvestments under the PRD policy. In other words, the same depreciation method may be represented by many different depreciation vectors by simply adding zeros at the end. This lack of uniqueness produces an ambiguity which we want to avoid. Hence, we require that the last element in the depreciation vector always be positive, i.e.,

$$(5) \qquad\qquad\qquad d_n > 0,$$

thus replacing $(.5, .3, .2, 0, 0, 0)$ by $(.5, .3, .2)$, $(0, 0, 1, 0, 0)$ by $(0, 0, 1)$, $(.5, 0, 0, .5, 0)$ by $(.5, 0, 0, .5)$, etc.[9]

2. Investment Ratio

Now let a_i be the ratio of the amount to be invested under the PRD policy at the *end* of Period i $(i = 0, 1, 2, \cdots)$ to the amount of the initial investment. Or alternatively, a_i represents the amount to be invested at the end of Period i when \$1 is invested initially. (We shall use both interpretations of a_i interchangeably not only for convenience of illustration but also for compactness and flexibility in interpreting results.) Then, it is easy to see that under the PRD policy stated in the previous section such a_i's may be written as:

$$(6) \qquad\qquad a_1 = d_1 a_0$$
$$a_2 = d_1 a_1 + d_2 a_0$$
$$a_3 = d_1 a_2 + d_2 a_1 + d_3 a_0$$
$$\cdots\cdots\cdots\cdots\cdots$$
$$\cdots\cdots\cdots\cdots\cdots$$

This may be simplified to:

$$(7) \quad a_k = d_1 a_{k-1} + d_2 a_{k-2} + \cdots + d_n a_{k-n} = \sum_{i=1}^{n} d_i a_{k-i}, \quad (k = 1, 2, \cdots, n)$$

where

$$(8) \qquad\qquad a_j \equiv 0 \quad \text{for} \quad j < 0 \quad \text{and} \quad a_0 \equiv 1.$$

We can then calculate a_k's for any k based on (7). For example, if

$$d = (d_1, d_2, d_3, d_4)$$

---

[9] Of course, it is possible to use $(.5, .5)$ to represent a depreciation method for assets with a two year life and $(.5, .5, 0, 0, 0)$ for assets with a five year life which are to be depreciated fully in the first two years. However, the life of the assets is irrelevant in deriving the PRD constant; hence, we shall use n to represent the period in which the asset is to be fully depreciated and not necessarily the life of the asset in the sense of its continuing use or retention by the company even after it has been fully depreciated.

then a_1 through a_6 are derived as follows:

(9) $\quad a_0 = 1$

$\quad a_1 = d_1$

$\quad a_2 = d_1^2 + d_2$

$\quad a_3 = d_1^3 + 2d_1d_2 + d_3$

$\quad a_4 = d_1^4 + 3d_1^2d_2 + 2d_1d_3 + d_2^2 + d_4$

$\quad a_5 = d_1^5 + 4d_1^3d_2 + 3d_1^2d_3 + 3d_1d_2^2 + 2d_1d_4 + 2d_2d_3$

$\quad a_6 = d_1^6 + 5d_1^4d_2 + 4d_1^3d_3 + 6d_1^2d_2^2 + 3d_1^2d_4 + 6d_1d_2d_3 + d_2^3 + 2d_2d_4 + d_3^2.$

As the simple example in (9) already suggests, the direct expression for a_k becomes very complicated for k large. We, therefore, direct our attention to the composition of book values of the fully depreciable assets in order to derive the PRD constant in the following analysis.

3. Book Values

Notice that since we set $a_0 \equiv 1$, the book value of the entire assets, denoted by b, will thereafter remain equal to unity, by virtue of the fact that new investment is always immediately undertaken in an amount equal to depreciation. On the other hand, the portion of b to be depreciated i periods later will change along with the age composition of the assets. Let us concentrate on this change in the age composition for a moment.

The example in Table 1 provides us with a convenient means to start our analysis. Suppose that, in this example, we partition the book value of the entire assets into a portion which is to be depreciated one period later and a portion which is to be depreciated two periods later. For example, at the end of Period 0 (after the initial investment was made), the portion to be depreciated one period later (Period 1) is .5 and the portion to be depreciated two periods later (Period 2) is also .5. However, at the end of Period 1 after the investment of $.5, the composition is changed to .75 and .25, with $.75 to be depreciated one period later (Period 2) and $.25 two periods later (Period 3). At the end of Period 2 after the investment of $.25, the composition is changed to .625 and .375; and so on.

Let us denote by b_{ki} the portion to be depreciated at the end of Period $(k + i)$ based on the age composition of the book value at the end of Period k after the depreciation and investment for this period have been made. In the above examples, $b_{01} = b_{02} = .5$; $b_{11} = .75$, $b_{12} = .25$; $b_{21} = .625$, $b_{22} = .375$, etc. Here, as mentioned earlier, we have

(10) $\qquad\qquad b_{k1} + b_{k2} + \cdots + b_{kn} = 1 \qquad (k = 0, 1, 2, \cdots),$

and, in particular,

(11) $\qquad\qquad\qquad b_{0i} = d_i \qquad\qquad (i = 1, 2, \cdots, n)$

by definition.

4. Depreciation Analysis

Now notice that b_{ki}, the portion to be depreciated in Period $k + i$ after the depreciation and investment in Period k, consists of the following two components:

(i) The fund provided for Period $k + i$ by the investments in Period $k - 1$ or earlier. This amount is given by $b_{k-1,i+1}$, i.e., the portion to be depreciated in Period $k - 1 + i + 1 = k + i$ after the depreciation and investment in Period $k - 1$.

(ii) The fund provided for Period $k + i$ by the investment in Period k. The investment in Period k is given by $b_{k-1,1}$, i.e., the portion to be depreciated in Period $k - 1 + 1 = k$ after the depreciation and investment in Period $k - 1$. Hence, the fund provided for Period $k + i$ by the investment in Period k is $b_{k-1,1} \cdot d_i$.

Therefore, we can express $b_{k,i}$ by the following relationship with $b_{k-1,1}$ and $b_{k-1,i+1}$,

$$(12) \qquad b_{k,i} = b_{k-1,i+1} + b_{k-1,1} \cdot d_i \qquad (k = 1, 2, \cdots),$$

where, by definition,

$$(13) \qquad b_{k-1,n+1} = 0 \qquad (k = 1, 2, \cdots),$$

since no funds are provided for Period $k + n$ by the investment in Period $k - 1$ or earlier. In the above example,

$$b_{21} = b_{12} + b_{11} \cdot d_1 = .25 + .75 \times .5 = .625$$
$$b_{22} = b_{13} + b_{11} \cdot d_2 = 0 + .75 \times .5 = .375.$$

5. Derivation of the PRD constant

We are now ready to derive the PRD constant based on (12). As we shall show later, unless the depreciation vector d is *cyclic*—e.g., $(0, .5, 0, .5)$, $(0, 0, 0, .7, 0, .3)$ —a case which results in reinvestments in every other year, we can derive a set of non-negative constants $b_1, b_2, \cdots, b_n$, where

$$(14) \qquad b_1 + b_2 + \cdots + b_n = 1,$$

with the property that for any given positive constant ϵ there exists a number N such that for all $i\ (= 1, 2, \cdots, n)$

$$(15) \qquad |b_{ki} - b_i| < \epsilon \quad \text{for all} \quad k \geqq N.$$

Thus as k becomes large the deviation between b_{ki} and b_i becomes smaller and smaller, which also means the deviation between b_{ki} and $b_{k-1,i}$ becomes smaller and smaller. Hence, in (12) we may substitute $b_{k,i+1}$ and $b_{k,1}$ for $b_{k-1,i+1}$ and $b_{k-1,1}$, respectively, for a large k, and in the limit we can rewrite the relationship in (12) as

$$(16) \qquad b_i = b_{i+1} + b_1 \cdot d_i \qquad (i = 1, 2, \cdots, n)$$

with $b_{n+1} \equiv 0$.

This expression (16) provides a clue for deriving the constant b_i's. First, for

$i = n$, we have

(17) $$b_n = b_1 \cdot d_n ,$$

since $b_{n+1} \equiv 0$. Now from this and (16) we derive an expression for $i = n - 1$ as follows:

(18) $$b_{n-1} = b_n + b_1 \cdot d_{n-1} = b_1 \cdot d_n + b_1 d_{n-1} .$$

Similarly, the expression for $i = n - 2$ is given from (16) and (18) as

(19) $$b_{n-2} = b_{n-1} + b_1 \cdot d_{n-2} = b_1 \cdot d_n + b_1 \cdot d_{n-1} + b_1 \cdot d_{n-2} .$$

In this manner the entire collection of b_i's can be written in terms of b_1 and the d_i's as follows:

(20)
$$b_n = b_1 \cdot d_n$$
$$b_{n-1} = b_1 \cdot (d_n + d_{n-1})$$
$$b_{n-2} = b_1 \cdot (d_n + d_{n-1} + d_{n-2})$$
$$\cdots\cdots\cdots\cdots\cdots\cdots\cdots\cdots$$
$$b_2 = b_1 \cdot (d_n + d_{n-1} + \cdots + d_2)$$
$$b_1 = b_1 \cdot (d_n + d_{n-1} + \cdots + d_2 + d_1).$$

Adding all the equations in (20), we have

(21) $b_1 + b_2 + \cdots + b_n$
$$= b_1[n \cdot d_n + (n - 1)d_{n-1} + (n - 2)d_{n-2} + \cdots + 2d_2 + d_1].$$

However, since $b_1 + b_2 + \cdots + b_n = 1$ as mentioned earlier, we have

(22) $$b_1[n \cdot d_n + (n - 1)d_{n-1} + \cdots + 2d_2 + d_1] = 1,$$

or

(23) $$b_1 = 1/(d_1 + 2d_2 + \cdots + (n - 1)d_{n-1} + nd_n).$$

Now notice that b_{k1} is the amount of depreciation for Period $k + 1$ which is equal to the investment to be made at the end of this period, i.e., a_{k+1}. Therefore, the fact that b_{k1} converges to b_1 implies that a_k also converges to b_1. Hence, the PRD constant is given by (23) which is the same as the expression (1) in Section I.[10]

---

[10] My colleague, Arthur Veinott, pointed out to me that the recursive relationship represented by (7) and the formula (1) as its solution are the same as the ones discussed in [4, pp. 285–286] in dealing with recurrent events in general. In fact, d_i may be modified and interpreted to mean the "probability" that the investment will be recovered for the first time in exactly i periods after the investment was made, which corresponds to f_i in [4]. We can, in fact, follow this route and via Markov analyses study the age composition and other physical aspects of depreciation. See [1, Chapter 17]. Also note that $\sum_{i=1}^{n} id_i$ in the formula (1) may be interpreted as the average length of time to recover the investment. See the notion of "mean recurrent time" in [4, p. 352] or "mean first passage time" in [13, pp. 78 ff.] and [12, pp. 411 ff.].

This leaves the condition on d for b_{ki}'s to converge to b_i's as the only remaining point to be discussed in order to complete the proof that the PRD constant is actually given by (1). This will be done in the next section.

III. Convergence

1. Markovian Description of Process

First of all, we simplify the expression (12) by using vector and matrix notations as follows: Let β_k be a row vector whose elements are b_{ki}'s, i.e.,

$$(24) \qquad \beta_k = (b_{k1}, b_{k2}, \cdots, b_{kn}) \qquad (k = 0, 1, 2, \cdots),$$

where

$$(25) \qquad \beta_0 \equiv (d_1, d_2, \cdots, d_n).$$

Also let D be the $n \times n$ matrix given by:

$$(26) \qquad D \equiv \left[\begin{array}{c} d \\ \hline I_{n-1} \mid 0 \end{array} \right] \equiv \left[\begin{array}{cccccc|c} d_1 & d_2 & d_3 & \cdots & d_{n-1} & d_n \\ \hline 1 & 0 & 0 & \cdots & 0 & 0 \\ 0 & 1 & 0 & \cdots & 0 & 0 \\ \cdot & \cdot & \cdot & \cdots & \cdot & \cdot \\ 0 & 0 & 0 & \cdots & 1 & 0 \end{array} \right].$$

Then, expression (12) can be written simply as

$$(27) \qquad \beta_k = \beta_{k-1} D \qquad (k = 1, 2, \cdots),$$

or

$$(28) \qquad \beta_k = \beta_{k-1}D = \beta_{k-2}D^2 = \beta_{k-3}D^3 = \cdots = \beta_0 D^k \qquad (k = 0, 1, 2, \cdots),$$

since D is a constant matrix.[11]

For the example in Table 1, we have

$$\beta_0 = (.5, .5)$$

and

$$D = \begin{bmatrix} .5 & .5 \\ 1 & 0 \end{bmatrix}.$$

Hence

$$\beta_1 = (.5, .5) \begin{bmatrix} .5 & .5 \\ 1 & 0 \end{bmatrix} = (.75, .25).$$

Similarly,

$$\beta_2 = (.75, .25) \begin{bmatrix} .5 & .5 \\ 1 & 0 \end{bmatrix} = (.625, .375),$$

[11] By D^k we mean $D \cdot D, \cdots D$ (k times).

which may also be obtained as

$$\beta_2 = (.5, .5) \begin{bmatrix} .5 & .5 \\ 1 & 0 \end{bmatrix}^2 = (.5, .5) \begin{bmatrix} .75 & .25 \\ .5 & .5 \end{bmatrix} = (.625, .375).$$

Notice that since the elements in the matrix D are all non-negative and the sum of elements in each row is always equal to one, D satisfies the requirement for Markov transition matrices.[12] Also note that β_k is a probability vector, i.e., all elements are non-negative and sum up to 1. Therefore, the process represented by (27) is Markovian and we can take advantage of the theorems[13] that have been developed in connection with Markov chains in deriving the conditions for the convergence of b_{ki}'s.

2. Regular and Cyclic Depreciation Vectors

Now let us classify depreciation vectors into *regular depreciation vectors* and *cyclic depreciation vectors* as follows. We call a depreciation vector regular if the greatest common denominator of all the indices i for which d_i's are positive is one, and cyclic otherwise. For example, $(0, 0, .5, 0, .4, .1)$ is regular since the greatest common denominator for all the indices for which d_i's are positive—viz, $i = 3, 5$ and 6—is 1, whereas $(0, 0, .5, 0, .0, .5)$ is cyclic since the greatest common denominator of all the indices for which d_i's are positive—viz, $i = 3$ and 6—is 3.

Then the conditions for the convergence of β_k may be stated in the following form.

Theorem 1: The series $\beta_0, \beta_1, \beta_2, \cdots$ is convergent if and only if d is a regular depreciation vector.

We shall prove this as follows: From the recursive relationship in (28), it is clear that the series of β_k's converges if and only if the series of $D^0, D^1, D^2, \cdots$ converges. However, D is a transition matrix for an *ergodic chain*, since the entry in the $(i + 1)$-st row and the i-th column of D is positive for $i = 1, 2, \cdots, n - 1$ and the entry in the 1-st row and the n-th column (d_n) is also positive by definition (5), so that it is therefore possible to "go from any state to any other state" in at most n steps. That is, for any i and j $(i, j = 1, 2, \cdots, n)$ b_{ki} affects $b_{k+l,j}$ for some positive integer $l \leq n$.

An ergodic transition matrix D is said to be *regular* if for some k the elements in D^k are all positive, and is said to be *cyclic* otherwise. It has been proved that

---

[12] Also note that D is essentially identical with the matrix used in representing a polynomial by the characteristic polynomial of the linear transformation defined by the matrix [6, p. 102].

[13] See, e.g., [12] and [13] for the theory of Markov chains. There have been a few interesting applications of Markov chain analyses to business problems. See [2] where an absorbing Markov chain is applied to the behavior of accounts receivable of a department store. Also an absorbing Markov chain is used in [17] to derive a depreciation vector when a survival probability is specified for each period of the asset life. Finally, see [9] where an application is made to machine repair and replacement problems by incorporating a payoff matrix [8] into the Markov chain analysis.

the series D^k's is convergent if D is regular, and not convergent if D is cyclic.[14] Therefore, all we need to show in order to prove the theorem is that D is regular if d is regular, and D is cyclic if d is cyclic.

Let $\{j_1, j_2, \cdots, j_l\}$ be the set of all subscripts such that $d_{j_i} (i = 1, 2, \cdots, l)$ is positive, and let J be the set of all integers j that are in the form

$$(29) \qquad\qquad j = r_1 j_1 + r_2 j_2 + \cdots + r_l j_l ,$$

where $r_i (i = 1, 2, \cdots, l)$ are all non-negative integers. Then it is easy to see that the investment in period j is positive $(a_j > 0)$ if and only if j is a member of J. For example, if $d = (0, 0, 0, .7, 0, .3)$, we have $j_1 = 4$ and $j_2 = 6$, hence J consists of integers 4, 6, 8, 10, 12, 14, $\cdots$.

Now let m be the greatest common denominator of $j_1, j_2, \cdots, j_l$. If $m = 1$, that is, if d is regular, it has been proved[15] that there exists an integer N such that every integer greater than or equal to N is a member of J. Hence, we have $a_k > 0$ for all $k \geq N$. This implies that there exists an integer M such that for all $k \geq M$ the elements of β_k are all positive, since $M = N + n$ clearly satisfies the requirement. However, $\beta_k = \beta_0 D^k = (1, 0, 0, \cdots, 0)D^{k+1}$ by (28) and by the fact that $\beta_0 \equiv (d_1, d_2, \cdots, d_n) = (1, 0, 0, \cdots, 0)D$, hence β_k is identical to the first row of D^{k+1}, which implies that the first row of D^{M+1} consists of positive elements only. Furthermore, all the elements immediately below the diagonal elements of $D^{kn+1} (k = 0, 1, 2, \cdots)$ are always positive for any d due to the cyclic pattern generated by the identity submatrix of D and d_n which is positive. Therefore, there exists an integer $K = kn + 1 \geq M + 1$ such that the elements in the first row of D^K and the elements immediately below the diagonal elements of D^K are all positive. From this, it is clear that the elements of D^{nK} are all positive; hence D is regular if $m = 1$, i.e., if d is regular.

On the other hand, if $m > 1$, J does not contain integers expressible as $rm + s$ where $r = 0, 1, 2, \cdots$ and $s = 1, 2, \cdots, m - 1$, which implies that for any N there exists $k \geq N$ such that k is not a member of J, i.e., $a_k = 0$. Since by definition a_k is equal to the first element of β_{k-1} which is identical to the first row of D^k, it follows that D^k contains at least one zero element in the first row and the first column for all $k = rm + s, r = 0, 1, 2, \cdots$ and $s = 1, 2, \cdots, m - 1$. This means that D^k contains at least one zero element for all $k = 0, 1, 2, \cdots$, since if the elements of D^k are all positive the elements of $D^{k+1}, D^{k+2}, \cdots$, are also necessarily all positive. Thus, D is cyclic if $m > 1$, i.e., if d is cyclic. This completes the proof of the theorem.[16]

3. Conversion of Cyclic to Regular Depreciation Vectors

If d is a cyclic depreciation vector, a new regular depreciation vector can be secured by eliminating all d_i's for $i = rm + s (r = 0, 1, 2, \cdots ; s = 1, 2, \cdots,$

---

[14] See, e.g., [13, pp. 69–71] and [12, pp. 399–402].

[15] See [13, pp. 6–7].

[16] My colleague, Takeshi Amemiya, showed me that the same result can also be obtained by solving the characteristic polynomial of D.

$m - 1$) from d and adjoining all remaining d_i's. Then the above analysis can be applied to the new D which is now regular and so then we only need to adjust the results of the analysis for the redundant periods that were eliminated. For example, a cyclic depreciation vector $(0, 0, .5, 0, 0, .5)$ can be converted into a regular depreciation vector $(.5, .5)$ in order to apply the analysis in the previous sections to this new vector, thereby obtaining the value $b_1 = \frac{2}{3}$ and $b_2 = \frac{1}{3}$ to which b_{k1} and b_{k2} converge, respectively. Then we can interpret the result to mean that the investment will be made in every three years by the amount approximately equal to $\frac{2}{3}$ of the initial investment in the long run.

4. Speed of Convergence

Finally, we shall briefly mention the speed of the convergence of a_k to the PRD constant. Let us denote by S the *smallest* integer such that, for a given positive ϵ we will have $|a_k - a| \leq \epsilon$ for all $k \geq S$. This information is very useful for managerial planning since it tells us when we can consider the periodic reinvestments to be stabilized, i.e., within an allowable range. Unfortunately, it is difficult to state S as a function of d and ϵ explicitly, although it is very simple to make a computer program which generates S when d and ϵ are supplied by using the recursive relationship (7).

However, if we know the value of s such that D^k contains no zero elements for all $k \geq s$ and the smallest element in D^s, then we can derive an S^* which *guarantees* that for a given positive ϵ, $|a_k - a| \leq \epsilon$ for all $k \geq S^*$, although S^* may not be the smallest among all such integers. We shall elaborate on this below.

Let α_k be a column vector whose elements are a_k, a_{k-1}, $\cdots$, a_{k-n+1}, where a_j is, as before, the amount of investment to be made at the end of Period j. Namely, we have

$$(30) \qquad \alpha_k = \begin{pmatrix} a_k \\ a_{k-1} \\ \vdots \\ a_{k-n+1} \end{pmatrix},$$

where $a_j \equiv 0$ for $j < 0$ and $a_0 \equiv 1$. Then the recursive relationship given in (7) may be written simply as:

$$(31) \qquad a_k = d\alpha_{k-1} \qquad\qquad (k = 1, 2, \cdots).$$

Or, by using the matrix D,

$$(32) \qquad \alpha_k = D\alpha_{k-1} \qquad\qquad (k = 1, 2, \cdots).$$

Next, let M_k and m_k be the largest and the smallest elements, respectively, in α_k, i.e., M_k is the largest a_j and m_k is the smallest a_j for $k - n + 1 \leq j \leq k$. Then, due to the averaging process represented by the first row of D, we have

$$(33) \qquad 1 = M_0 \geq M_1 \geq M_2 \cdots,$$

$$0 = m_0 \leq m_1 \leq m_2 \cdots,$$

thus,

$$(34) \qquad 1 = M_0 - m_0 \geqq M_1 - m_1 \geqq M_2 - m_2 \cdots \geqq 0.$$

Now, as we observed earlier, if d is regular, zero elements in D^k are gradually eliminated as k increases and eventually we will have a D^k which contains no zero elements. Let s be an integer for which D^s contains no zero elements, and let d_Δ be the smallest element in D^s. We can then state the bound for $M_{k+s} - m_{k+s}$ in terms of $M_k - m_k$ by using the theorem proved in [13], p. 70, or [12], p. 400, as follows:

$$(35) \qquad M_{k+s} \leqq (1 - d_\Delta)M_k + d_\Delta m_k,$$
$$m_{k+s} \geqq d_\Delta M_k + (1 - d_\Delta)m_k.$$

Hence,

$$(36) \qquad M_{k+s} - m_{k+s} \leqq (1 - 2d_\Delta)(M_k - m_k).$$

By setting $k = 0$ and using the recursive relationship, $\alpha_{2s} = D^s \alpha_s = D^{2s} \alpha_0$, we have

$$(37) \qquad M_{ls} - m_{ls} \leqq (1 - 2d_\Delta)^l,$$

since $M_0 - m_0 = 1$. From this we can derive[17] for a given ϵ the value of S^* which guarantees that $|a_k - a| \leqq \epsilon$ for all $k \geqq S^*$.

References

1. CHURCHMAN, C. W., ACKOFF, R. L. AND ARNOFF, E. L. *Introduction to Operations Research*, New York: John Wiley & Sons, Inc., 1957.
2. CYERT, R. M., DAVIDSON, H. J. AND THOMPSON, G. L., "Estimation of the Allowance for Doubtful Accounts by Markov Chains." *Management Science*, Vol. 8, No. 3 (April 1962), pp. 287–303.
3. EISNER, ROBERT AND STROTZ, ROBERT H., "Determinant of Business Investment," Research Study Number 2 in *Impacts of Monetary Policy*, prepared for the Commission on Money and Credit, Englewood Cliffs, N. J.: Prentice-Hall, 1963, pp. 60–333.
4. FELLER, W., *An Introduction to Probability Theory and Its Applications*, Vol. 1, 2nd ed. New York: John Wiley & Sons, Inc., 1957.
5. GRANT, E. L. AND NORTON, P. T., JR., *Depreciation*, Revised Print, New York: The Ronald Press Co., 1955.
6. HALMOS, P. R., *Finite-Dimensional Vector Spaces*, 2nd edition, Princeton, N. J.: D. Van Nostrand Co., Inc., 1958.
7. HICKS, J. R., *Value and Capital*, 2nd edition, Oxford: The Clarendon Press, 1946.
8. HOWARD, R. A., *Dynamic Programming and Markov Processes*, New York: John Wiley & Sons, Inc., 1960.
9. IJIRI, Y., "An Application of Finite Markov Processes to Machine Repair and Replacement Problems," unpublished term paper, Graduate School of Industrial Administration, Carnegie Institute of Technology, May 1961.
10. —— AND ROBICHEK, A. A., "Accounting Measurement and Rates of Return," Working Paper No. 30, Graduate School of Business, Stanford University, February, 1965.

[17] This also proves Theorem 1 which we used earlier by supposing that if there exist k such that D^k contains only positive elements, i.e., if D is regular, then the series of D^k's is convergent.

11. JORGENSON, DALE W., "Capital Theory and Investment Behavior," *American Economic Review*, May 1963, pp. 247-259.

12. KEMENY, J. G., MIRKIL, H., SNELL, J. L. AND THOMPSON, G. L., *Finite Mathematical Structures*, Englewood Cliffs, N. J.: Prentice-Hall, Inc., 1959.

13. —— AND SNELL, J. L., *Finite Markov Chains*, Princeton, N. J.: D. Van Nostrand Company, Inc., 1960.

14. LAYA, J. C., *A Cash Flow Model and the Rate of Return: The Effect of Price Level Change and Other Factors on Book-Yield*, Ph.D. Dissertation, Graduate School of Business, Stanford University, Stanford, California, August, 1965.

15. SAMETZ, ARNOLD W., "Trends in the Volume and Composition of Equity Finance," *Journal of Finance*, Vol. 19, No. 3 (September 1964), pp. 450-469.

16. SHAPIRO, ELI AND WHITE, WILLIAM L., "Patterns of Business Financing: Some Comments," *Journal of Finance*, Vol. 20, No. 4 (December 1965), pp. 693-707.

17. ZANNETOS, Z. S., "Markov Chain Depreciation Model and Survivor Curve Depreciation," *Metroeconomica*, Vol. 15, 1963, pp. 155-66.

Depreciation and Capital Maintenance: Hicks's Views

The Accounting Historians Journal
Vol. 9, No. 1
Spring 1982

Richard P. Brief
NEW YORK UNIVERSITY

HICKS ON ACCOUNTING

Abstract: Whenever income and capital maintenance concepts are discussed at the conceptual level, a reference to Hicks is likely to be found. These references are misleading since Hicks himself believed that the proper basis of valuation in the financial statements of a firm is historical cost. He also argued that accountants should not make price-level adjustments. Hicks' views on accounting, which are scattered in his writings over a period of 35 years, are reviewed in this paper.

J. R. Hicks viewed accounting from a statistician's perspective and he emphasized the need for objective accounting data. Hicks also was concerned with the principles of account classification which he once called "the canons of orderliness."[1] The need for objectivity and order, together with a very strict interpretation of how accounts should be framed to monitor management, led this 1972 Nobel Prize economist to defend the practice of valuing assets at historical cost and to argue, as a corollary, that accountants should not make price-level adjustments.

This short description of Hicks on accounting differs sharply from the popular view. The misconception is due to the pervasive influence on accounting thought of Hicks' definition of a man's income as "the maximum value which he can consume during a week, and still expect to be as well off at the end of the week as he was at the beginning."[2] This income concept was introduced into accounting literature by Alexander in 1950 and, by the early 1960s, Zeff reported that the definition "recurs with remarkable frequency in economic and (especially) accounting writings."[3] Today, whenever the income concept is discussed at a conceptual level, a reference to Hicks is likely to be found.

Hicks himself warned that income and related concepts are "bad tools, which break in your hands."[4] However, with few exceptions, most theorists have not only ignored this admonition, but they also have overlooked other work by Hicks which is more directly related to accounting practice.

Hicks' interest in accounting was signalled in 1942 with the publication of *The Social Framework*. This book is a text on "Social

Accounting," which the author contends should be a first course in economics. Social Accounting is defined as "the accounting of the whole community or nation, just as Private Accounting is the accounting of the individual firm."[5] In the second edition (1952), Hicks commented that "when I wrote in 1942, I knew very little of the practice of accountants, and I am afraid professional accountants who read my book were often bothered by needless unorthodoxy in the use of terms." Among others, F. Sewell Bray is thanked for some "excellent coaching in accounting usages."[6]

Interesting ideas on accounting also are contained in *The Problem of Budgetary Reform* (1948) and noteworthy opinions are expressed in a review of a book by Bray that was published in the *Economic Journal,* also in 1948. Other works which have special significance for accountants include: "The Measurement of Capital" (1969), *Capital and Time* (1973) and "Capital Controversies: Ancient and Modern" (1974).

The purpose of this paper is to review Hicks' main thoughts on accounting practice. They concern the need for objectivity, the principles of account classification and the traditional questions of depreciation and asset valuation. These ideas, which are scattered in books and articles written over a period of 35 years, make it evident that Hicks' actual influence on accounting was the opposite of what he intended.

Objectivity of Accounting Measurements

In a 1948 review of a book by Bray, Hicks emphasized the need for objectivity and then defended the cost principle. He used the following argument to reach this conclusion.

The function of accounting, on the lowest level, is to make a record of business transactions. On a higher level, accountants devise a means to distill from the record of transactions summaries which "enable the meaning of the record to be grasped, as well as it can be grasped."[7] These summaries should be constructed using rules which do not introduce outside elements. But since no set of rules will produce summaries "which are equally meaningful on all occasions,"[8] the user of financial statements must exercise judgment in interpreting them.

Although the interpretation of financial statements is considered the highest part of the accountant's function, it should not get "mixed up" with the lower part because the accounts must be "as objective as possible."[9] "This demand is so exactly parallel to the demand that statisticians have had to make of their investigators . . . that its sense is readily acceptable, once we see it . . . ."[10]

The consequence of this position led to the conclusion that the justification for valuation at historical cost is "simply that the original purchase price is the one objective valuation of the asset . . . . Any other valuation must be a matter of judgment, and therefore belongs to the stage of interpretation, not to the stage of the compilation of the basic summaries."[11]

The need for objectivity is the central issue in accounting and it has two further implications. First,

> It reproves those economists who have demanded from the basic accounts information which those accounts cannot properly give. And it also reproves those accounting practices (such as the practice of valuing stocks at cost or market price, *whichever is the lower)* which edge a little interpretation into the work of summarising, and therefore diminish the objectivity of the basic tables.[12]

Second,

> Every statistical table needs to be annotated, and in a similar way the full meaning of an accounting statement can only be expected to emerge in the accompanying report. But the accountant's report goes to the directors, while his figures go to the shareholders; he has thus some public obligation to pack into his figures the maximum of information, even if he can only do this, within the limits prescribed, by some sacrifice of objectivity. How ought this difficulty to be got over? Should it be laid down that companies must publish an audited report as well as audited accounts? Or would this make the accountant, more than ever, master of the destinies of us all?[13]

Ordering of Accounts

In 1948 Hicks also wrote a short monograph, *The Problem of Budgetary Reform,* and it dealt with the classification problem. The importance he accorded this problem is evident in the initial discussion of the accounting and economic aspects of the definition of a budget surplus. The accounting aspect concerned maintaining order in government accounts and the economic dimension related to the relationship between government accounts and the rest of the economy. Although Hicks explicitly states that he writes as an economist, he strongly contended that the accounting aspect is the most important consideration.

> If we can agree upon the principles by which the government's accounts should be ordered, the accounts of National Income and Expenditure could be adjusted to fit; but if we start from the other end, beginning from the economic requirements, we may easily be endeavoring to fit the government accounts into a framework with which it is difficult, and from their own point of view may be undesirable, for them to conform.[14]

For Hicks the principles of account classification had to do with problems like whether a distinction ought to be made between current and capital items. The main issue was not a matter of economics but rather a question of purpose. Different systems of account classification were needed for different purposes. Hicks repeatedly stressed the point of view that purpose is the critical determinant of accounting practice.

The possibility of putting government finances on a business accounting basis also was mentioned, but this "drastic" remedy was rejected even though

> It would provide a new and authoritative set of canons for orderliness; it would enlist the experience and influence of the accountancy profession in maintaining those standards; and it would facilitate the integration of the government's accounts with those of the rest of the economy in national income calculations, with all that implies for the smooth working of rational employment policy.[15]

A similar statement was made in the second edition of *The Social Framework* (1952) and it was repeated in the fourth edition (1971).

> The first thing that has to be done is to prepare the bricks out of which the structure is to be built, by constructing a standard set of accounts for the individual units out of which the national economy is composed. Much of this task has already been performed by professional accountants, and we can draw heavily upon their work at this stage of the argument.[16]

Hicks often referred to the accountant's positive contributions as well as his influential position. He even once discussed the influence of accounting practice on the history of economic thought.[17]

Depreciation Accounting

In 1939 Hicks and Hicks remarked that the calculation of depreciation presents "awkward problems."[18] This statement was often repeated. Hicks also called depreciation an "artificial" item that does not correspond to actual cash payments. It is "less solid" than other items in an income statement and there is "some room for judgment" about the precise sum to be put down, though accountants are generally guided by conventional rules about this "most uncertain" item.[19] Over a quarter of a century later similar comments are made.

> There are items, of which depreciation and stock accumulation are the most important, which do not reflect actual transactions but are estimates (by the accountant, not by the statistician) of changes in the value of assets which have not, or not yet, been sold.[20]

Exactly the same views were expressed in *Capital and Time* where the meaning of these estimates was elaborated on.

> These are estimates in a different sense from that previously mentioned. They are not statistician's estimates of a true figure, which happens to be unavailable; there is no sure figure to which they correspond. They are estimates that are relative to a purpose; for different purposes they may be made in different ways.[21]

Hicks never criticized the accountant's method of calculating depreciation. He simply observed that there was no firm economic solution to the depreciation problem and that the methods developed by accountants as soon as they were confronted with the problem "were probably what they had to do . . . . It is what they still do, even in this day."[22]

All English editions of *The Social Framework* (1942, 1952, 1960, and 1971) and the American editions (1945 and 1955) contained an Appendix on depreciation. The main point was that different people might estimate depreciation in different ways and the same person might have different estimates for different purposes. Again, the focus is on purpose.

Two particular estimates for the depreciation of a firms' fixed capital are discussed. The first is for the purpose of determining profits available for dividends and the second is made for the purpose of taxation. Conservative principles govern the calculation on which dividends are paid, whereas tax laws are based on a concept of fairness.

Hicks also pointed out that these depreciation estimates are not satisfactory for the purpose of determining national income when prices are unstable. Under inflationary conditions it is the task of the economists to work out methods to compute real depreciation. However, "although it is necessary, in the interest of fairness, to go back to the original purchase price (for that is firm ground, not somebody's guess), to do this is not economically satisfactory."[23] But Hicks never suggested that accountants ought to make this computation.

Asset Valuation

The idea that "the fixed capital used by a manufacturing firm may have half a dozen different values that can be plausibly put upon it"[24] often was repeated. Purpose was paramount.

> The measurement of capital is one of the nastiest jobs
> that economists have set to statisticians. Finding that it is
> so nasty, the working satistician very naturally asks for
> guidance. Will the economist please explain just what it is
> that he wants?[25]

The 1969 paper explicitly analyzes the accountant's practice of valuing assets at original cost and two reasons are given for making corrections to this valuation—price-level changes and technical progress.

It is significant that Hicks did not think that accountants should be the ones to make price-level adjustments. He concluded that the task of making price-level adjustments would be a formidable one and implied that on a cost-benefit basis, it would not be worthwhile to have accountants correct the accounts for inflation. Hicks put it this way.

> To correct balance-sheets of companies, one by one, so
> as to make them conform to the principle I have been out-
> lining would obviously be a formidable undertaking. One
> could not expect that it should be done by the account-
> ants, the purpose of whose calculations (as we have seen)
> is quite different.[26]

What was the purpose of the accountant's calculations? Hicks believed that "the first object of commercial accounting is to watch over the capital of the business . . . ."[27] Thus, accountants could not be expected to make price-level adjustments because

> If one asks why the company needs a balance-sheet the answer must surely be given in terms of its obligation to its shareholders and to other creditors. It has to show, periodically, what it has done with their money.[28]

This emphasis on stewardship led to the conclusion that

> The balance-sheet of the company, as explained, is designed to show shareholders what has been done with their money. If it is this which has to be shown, the original cost of the actual assets held is the magnitude that is relevant.[29]

The other reason for adjusting original cost is technical progress. However, from an accounting viewpoint, technical progress is part of obsolescence and is taken into account in choosing the method of depreciation. But from an economic viewpoint, Hicks argues the accountant is too pessimistic because "he does not allow (and from his own point of view is quite right not to allow) for the increase in the productivity of new investments that come from technical progress."[30] Therefore, the accountant's measure of depreciation normally overstates the gross investment needed to maintain capital since real capital increases faster than it appears to do on financial statements since the reinvestment of depreciation allowances is made on continually more favorable terms.

There is no ambiguity about Hicks thoughts on the accountant's responsibility for asset valuation. Financial statements should be based on historical costs and adjustments for price-level changes should be made by the user, not the accountant.

Concluding Comment

At the 1969 meeting of the International Statistical Institute where Hicks presented the paper, "The Measurement of Capital," another paper on the same subject was given. The opening remarks are jocular, but the message captures Hicks' attitude towards the valuation problem which accountants face.

> Without the threat of thumbscrew, and indeed with no urging at all, an economist may often be found to declare that his idea of a measure for a stock of capital is to equate it to the present discounted value of the future stream of

earnings that the stock of capital will generate. This is so inherently unmeasurable that it will amuse a statistician until he perceives that the suggestion is offered somewhat more than half-seriously.[31]

When it came to practical problems of measurement, Hicks sought to avoid the kind of definitions that are more "at home in those simplified models beloved of economic theorists" than in a world of "flesh and blood."[32]

Hicks' emphasis on the need for objectivity and his conception of the meaning of stewardship led him to defend the traditional practice of valuing assets at original cost less depreciation and to argue that accountants should not make price-level adjustments. These views will surprise those (and there are many) who have used the Hicksian income concept to develop a conceptual framework in accounting.

POSTSCRIPT

After this article went into production the author received a letter from Sir John Hicks. Although Hicks indicated that he did not have time to do more than glance at a copy of the article, he commented that "I had no idea when I wrote that chapter in *Value and Capital* that it would be taken up by accountants; and The Social Accounts as I first envisaged them, were a pure economist's construction. By the time of the later editions of that book [1952, 1960, and 1971] I had this further experience, and the same applies to all my later writings." The experience Hicks is making reference to is a 1949 meeting (which was referred to in the 2nd edition of *The Social Framework*) with, among others, Richard Stone and F. Sewell Bray. This meeting led to the publication of *Some Accounting Terms and Concepts* (Cambridge: Cambridge University Press, 1951). I have not seen this book. Hicks then went on to say that "it is perhaps in the IIPF paper ["The Concept of Income in Relation to Taxation and to Business Management," Proceedings of the 35th Congress of the International Institute of Public Finance, Taormina, 10-14 September, 1979, Detroit, Michigan 1981] which I sent to you that this becomes clearest."

The 1979 paper compares the accountant's depreciation with "true" depreciation. Its basic conclusions are consistent with his earlier work. Hicks once again stressed that "The accountant's conventional way of measuring income, though (as we have seen) it has elements in it that are arbitrary, is largely based upon actual transactions . . . . It would not be an improvement to replace this relatively firm assessment by one that in practice must be even more, even much more arbitrary." He also emphasized that "when it is proposed to make corrections to the conventional allowances, to adjust to inflation, the question of whether the conventional allowances were appropriate, even in absence of inflation, is bound to be raised. So one comes back to the subjective assessments— how much can be *safely* taken out of the business—from which the accountant's procedure had been thought to be an escape."

Both the 1974 paper delivered at the annual meetings of the American Economic Association and the paper given at the IIPF meeting in 1979 should interest the accounting historian for another reason. In these papers, Hicks directly addresses questions involving the history of accounting. In addition, in his letter Hicks also remarked that "I see that the journal in which you are publishing is on the history of accounting. I wonder if it is known to your friends that one of the most remarkable contributions of accounting to civilization is the beginning of the year on the first of January. The Florentine merchants in the fifteenth century found this sense of orderliness (as you rightly call it) vexed by the practice then and in other places long after common of beginning on March 25. They wanted to have their yearly accounts made up of a tidy number of monthly accounts, so they looked for a feast of the Church which fell on the first day of the month, and found it on the first of January." The reference Hicks provides is to a French economic historian, Yves Renouard, who in 1969 published a book of essays which I have not yet seen.

I am grateful to Sir John Hicks for his thoughtful comments. I also thank Professor Edward Stamp for suggesting that I communicate with Hicks.

FOOTNOTES

[1]*The Problem of Budgetary Reform*, p. 14.
[2]*Value and Capital*, p. 172.
[3]Zeff, p. 620.
[4]*Value and Capital*, p. 177.
[5]*The Social Framework* (1942), p. vi.
[6]*The Social Framework* (1952), p. vii.
[7]Review, p. 562.
[8]Review, p. 563.
[9]Review, p. 563.
[10]Review, p. 563.
[11]Review, p. 563.
[12]Review, pp. 563-564.

[13]Review, p. 564.
[14]*The Problem of Budgetary Reform,* p. 6.
[15]*The Problem of Budgetary Reform,* p. 14.
[16]*The Social Framework* (1952), p. 224; (1971), p. 260.
[17]"Capital Controversies: Ancient and Modern," p. 310.
[18]"Public Finance in National Income," p. 147.
[19]*The Social Framework* (1942), p. 225.
[20]"The Measurement of Capital," p. 254.
[21]Capital and Time, p. 155.
[22]"Capital Controversies: Ancient and Modern," p. 312.
[23]*The Social Framework* (1942), p. 275. A similar statement is made in the fourth edition (1971), pp. 301-302.
[24]*The Social Framework* (1942), p. 102.
[25]"The Measurement of Capital," p. 254.
[26]"The Measurement of Capital," p. 259.
[27]*The Problem of Budgetary Reform,* p. 15.
[28]"The Measurement of Capital," p. 253.
[29]"The Measurement of Capital," p. 258.
[30]"The Measurement of Capital," p. 261.
[31]Evans, "Some Comments on Measures of Changes in Capital Stock Aggregates," p. 265.
[32]"Maintaining Capital Intact: A Further Suggestion," pp. 132-133.

BIBLIOGRAPHY

Alexander, Sydney S. "Income Measurement in a Dynamic Economy." *Five Monographs on Business Income.* American Institute of Accountants, 1950; reprinted, Lawrence, Kansas: Scholars Book Co., 1973, pp. 1-96.
Evans, W. Duane. "Some Comments on Measures of Changes in Capital Stock Aggregates." *Proceedings of the 37th Session.* International Statistical Institute (London 1969). Vol. XLIII, Book 1, pp. 265-275.
Hicks, J. R. *Value and Capital.* Oxford: Clarendon Press, 1939.
_____ and Hicks, U. K. "Public Finance in National Income." *Review of Economic Studies* (February 1939), pp. 147-155.
_____. *The Social Framework.* Oxford: Clarendon Press, 1942.
_____. "Maintaining Capital Intact: A Further Suggestion." *Economica* (May 1942), pp. 174-179 in Parker, R. H. and Harcourt, G. C., eds. *Readings in the Concept and Measurement of Income.* Cambridge, University Press, 1969, pp. 132-138.
_____ and Hart, A. G. *The Social Framework of the American Economy.* New York: Oxford University Press, 1945.
_____. Review of *Precision and Design in Accountancy* (London: Gee, 1947), by F. Sewell Bray. *Economic Journal* (December 1948), pp. 562-564.
_____. *The Problem of Budgetary Reform.* Oxford: Clarendon Press, 1948.
_____. *The Social Framework.* 2nd ed. Oxford: Clarendon Press, 1952.
_____. Hart, A. G. and Ford, J. W. *The Social Framework of the American Economy.* 2nd ed. New York: Oxford University Press, 1955.
_____. *The Social Framework.* 3rd ed. Oxford: Clarendon Press, 1960.
_____. "The Measurement of Capital." *Proceedings of the 37th Session,* International Statistical Institute (London, 1969), Vol. XLIII, Book 1, pp. 253-263.

_____. *The Social Framework.* 4th ed. Oxford: Clarendon Press, 1971.

_____. *Capital and Time.* Oxford: Clarendon Press, 1973.

_____. "Capital Controversies: Ancient and Modern." *American Economic Review* (May 1974), pp. 307-319.

Zeff, Stephen A. "Replacement Cost: Member of the Family, Welcome Guest, or Intruder?" *The Accounting Review* (October 1962), pp. 611-625.

THE MEASUREMENT OF CAPITAL

By

JOHN RICHARD HICKS

All Souls College, Oxford

THE measurement of capital is one of the nastiest jobs that economists have set to statisticians. Finding that it is so nasty, the working statistician very naturally asks for guidance. Will the economist please explain just what it is that he wants? This is a question that calls for some heart-searching on the part of the economist. He has to ask himself very seriously: what is the point of putting the problem? Do we really need a measure of capital? If so, what for? I cannot avoid a discussion of these questions, though you may say that they are economist's not statistician's questions.

Let us remind ourselves how the matter arose. At the time, long ago, when statisticians first endeavoured to estimate national wealth, the distinction between income and capital was not at all clear. So it had to be insisted, in order to clarify it, that there were two things that might be measured, not one. Wealth has two aspects, stock and flow; without a clear idea of the stock concept, the flow concept itself is hard to define. The problem of measuring capital had to be envisaged, if only in order that there should be some clarification in the concept of income.

Familiarity with the form of accounts that is proper for a limited liability company was then a great help. The company has a balance-sheet, which is a statement of its wealth at a point of time, as well as running accounts, which are statements of its activities over a period. It has several running accounts, as conventionally expressed (trading accounts, profit and loss account, and so on) but these could easily be amalgamated into a single running account, if it was desired to do so. The balance-sheet is something different, which could not be amalgamated with them.

Once this distinction was appreciated, it was easy (and natural) to conceive of the social, or national, accounting problem in similar terms. There should, on the one hand, be an account of the national income, to be divided into separate running accounts in whatever way was thought to be convenient; and on the other hand there should be an account of the national capital, a national balance-sheet. So it seemed, by analogy; but it became evident, on closer examination, that the analogy is by no means exact.

If one asks why the company needs a balance-sheet, the answer must surely be given in terms of its obligations to its shareholders and to other creditors. It has to show, periodically, what it has done with their money. Thus it is that a company which does no trading must still have a balance-sheet; it must have a balance-sheet because of its legal position. Now though occasions arise (one of them may be found in the situation of the United Kingdom at the present time) in which a nation has got itself into a position, with respect to its foreign creditors, that is not altogether dissimilar, it is still under no legal obligation to present a balance-sheet. A national balance-sheet, or estimate of the national capital, such as may be prepared by a statistician, must inevitably have a different purpose.

It is characteristic of a running account, of whatever type, that most (though not all) of the items that enter into it are records of actual transactions. When an article is sold, money passes hands; so the value of the article is expressed in money terms, by buyer and seller, in the same way. When money is lent, currently, the same occurs. Thus if it were the case that all entities within an economy (all private individuals, as well as all traders) kept proper running accounts, and if those accounts contained nothing but transactions items, it would be possible for a national running account to be prepared from them by a purely arithmetical process. Many of the accounts which he would need for this purpose are, of course, not available to the national income statistician; he has to estimate them. He can nevertheless use the identity between buying and selling value as a means of estimating the accounts that are missing; not completely, but quite often to the full extent of his need to estimate them.

Even in the case of running accounts, things are not always so simple. There are items, of which depreciation and stock accumulation are the most important, which do not reflect actual transactions, but are estimates (by the accountant, not by the statistician) of changes in the value of assets which have not, or not yet, been sold. These, it is true, present no arithmetical difficulty (so long as the accounts are available). The depreciation allowances in the accounts of an individual firm appear in the accounts of that firm only, so that they can appear in a national account just by being added up. The significance of the total thus arrived at remains however highly questionable. Estimates of depreciation (or 'capital consumption') may evidently be made in different ways for different purposes. A firm may change its policy about them. The method of estimation which is suitable for the firm's own purposes is very generally different from that which is suitable (or indeed permitted) for other purposes, tax purposes or insurance purposes; what is permitted for tax purposes is itself liable to change. This is of course the basic reason why it has become customary to express the national accounts in terms of Gross National Product (before deduction for depreciation), so as to clear them of contamination with the 'arbitrary' depreciation item; though it should be noticed that even with GNP there remains another 'arbitrary' element in stock accumulation. Policies about its valuation may differ: FIFO and LIFO and so on.

What in the case of running accounts is a complication, that can thus to some extent be avoided, in the case of the balance-sheet is central and unavoidable. The assets, the possession of which is recorded in a balance-sheet, are assets that are held, not goods that are sold. They may be sold, when the time comes, but they are not being sold at the date to which the balance-sheet refers. The values which are set upon them, not indeed in all cases but in very many cases, are thus by their nature estimated values, estimates which, as before, may be made in different ways for different purposes.

There are indeed some cases, some important cases, in which this obstacle to objective valuation can in a sense be overcome. Actual cash, or near-cash, on the one side; liabilities payable on demand, or in the near future, on the other; everyone would grant that these are objectively assessable. They are even objective enough to appear at the same figure, or nearly the same figure, in the balance-sheets of debtor and creditor; thus the bank deposit, which is an asset of the depositor, appears as a corresponding entry among the liabilities of the bank. The one account matches the other, as it does, in the case of running accounts, when goods are sold. In this limited field, accordingly, consolidation is easy.

Take the case of marketable securities, securities in which there is an active market. Here there is, at least, a price, an objective price. It may vary very much from day to day, but 'at a point of time', the time to which the balance-sheet refers, it exists. The value which is set upon a share by the market is, in strictness, the valuation which is put upon it by the marginal seller and buyer; those who do not sell, but hold, must (presumably) set a higher value upon it. Yet there is a significant sense in which the market value of the share is the value of the share to the shareholder, to any share-holder; for it is the realizable value, the value at which he could sell, if he desired to do so. What is so troublesome is that it is not the value of the share to the company, to the company of which it is the liability. It is not the value which appears (even by implication) in the company's own balance-sheet. The basic reason for this discrepancy is the legal character of the company balance-sheet; reducible, in the end, to the legal provision which, for such good cause, prevents the company from trading in its own shares.

The existence of this discrepancy is a fundamental obstacle to the construction of an all-purpose national balance-sheet; but it does not imply that the valuation of securities at market value, at shareholders' value, is a method that should always be rejected. There is indeed one purpose of measuring capital for which it is beyond question appropriate. If our interest is in the *distribution* of capital, the distribution of personal property, we need only be concerned with the personal sector of the economy; and it is perfectly appropriate to take as the value of a person's property the value at which it could be *realized*. There are, after all, two things which may be done with a piece of property: it may be enjoyed, deriving an income from it ('psyche' or other), or it may be sold. The advantage of having property that can be sold to meet emergencies is an aspect of distributional inequality which could not be taken into account in a statement of the distribution of income (however comprehensive). It is precisely the thing which is shown up by a measurement of personal capital realizable value.

Wealth that is held in the form of marketable securities has a realizable value which is easily ascertainable; but even in the assessment of personal property there is a problem of wealth that is held in other forms. The values of land and houses, in particular, must (almost) always be estimated; and then there is a question of just what it is that is being estimated. It cannot be expected that they can be sold at a 'reasonable' price the moment that a decision to sell them is taken; some delay must be permitted; what delay? The statistician is in fact at the mercy of the public authority, with the provisions that it makes (from time to time) on such matters as the disposal of property for death duties; there is no clear economic criterion. It is nevertheless clear, in principle, what we should be trying to measure. The value of capital, from the point of view of personal distribution, is its realizable value.

Land and houses are nowadays the principal forms of physical property which are held directly by private persons, as private persons; they have a market, a very imperfect market it may be, but a market all the same. But what is to be done with those things that do not have a market, or hardly have a market? This is surely the case with the greater part of the assets of productive industry.

The issue arises, it will be noticed, at least to some extent, even while we are concerned with the distribution of property, in the case, the still not negligible case, of the unincorporated business. Its assets are directly owned, and they are pieces of productive equipment.

It is impossible to admit that the value of a piece of industrial equipment is the price at which it could be realized. If it were removed from the environment in which it is used, it would (in many cases) be nearly useless; it would have to be 'sold for scrap'. Its value, when used as it is being used, must be greater than that. But how is that value to be assessed?

If we stick to the realizable value criterion, there is only one way out. We must say that this is a case in which the value of an asset is put up by the existence of other assets which go with it, which are complementary with it; so that although it is impossible to divide up the value of the complex into the values of the individual assets which compose it, the complex itself, the firm in working order, can still be valued. The value of the unincorporated business is the value at which (on a still more imperfect market, no doubt, even than the land market) it could be sold.

But then, having gone so far, should we not go further? What precisely is the difference that is made by incorporation? Why cannot we say that the value of a company's shares, as revealed on the market, is the value of the company's assets (or, more strictly, of its net assets) in the same sense?

When the unincorporated business is sold, it is 'taken over'. A 'take-over bid', for an incorporated company, is not very different from the sale of an unincorporated business, and may indeed be reckoned as registering the value of the business, in much the same sense. Normally, however, the shares will stand at a price which is greater than that at which any bidder would bid for them, which is why no bid takes place. There are doubtless many reasons for this excess of the market price over the highest potential 'bid' price; it has several components. One of them, surely, is a pure liquidity premium; the advantage to the holder of having a highly marketable asset.

It is, I think, this liquidity element (which is another way of saying this speculative element) in the valuation of shares on the market that makes one so reluctant to admit that the market value of its shares is *the* value of the company, or of its capital. If we could admit it, it would indeed have great convenience. For then, apart from complications, such as the existence of nationalized industries and public property, into which I shall not enter, we could regard the total of private property, estimated at its realizable market value, as being equal to the total of the national capital. We could then construct a balance-sheet of the national capital on a consistent basis. We should have to doctor the accounts of companies, so as to bring out their net worth (assets minus liabilities, including shares, valued at market value) to be zero. There must, as we have seen, be some doctoring if we are to get a consistent balance-sheet. It is perfectly possible to construct (or to estimate) a balance-sheet on this basis, and it may be useful to construct it, at least for pedagogic purposes. (I have used it myself for such purposes, in Hicks (1952). In a revised third edition, in 1960, under the influence of Professor E. V. Morgan's balance-sheet, constructed on the opposite principle, to which I shall be coming, I am afraid I wobbled.) But except in relation to the distribution of property, where, as we have seen, it belongs, it is not of much use.

Its most evident defect is its utter inability to generate a meaningful concept of capital accumulation, for society as a whole, *in real terms*. How can we reduce it sensibly, the pieces of paper of which it is (apparently) so largely composed, into real terms? They are being valued at realizable value, which is significant only as being what they would fetch when sold (in sufficiently small amounts, of course, not to be too upsetting to their prices). The advantage that is being measured is that which would be got if they were sold, and the proceeds spent, presumably upon consumption. So, on this line of thought, their value should be put into real terms by deflating it with respect to a consumption price-index. But what could that signify?

The current values of securities are indeed affected by consumption good prices, but not just by current consumption good prices; they are also affected by what such prices are expected to be in the future. An expectation of inflation is likely to raise the market values of securities, relatively to current consumption good prices; thus on this criterion it will raise the national capital *in real terms*. Though this has a meaning in respect to the probability of dis-saving, it is (in general) a very inconvenient thing to have to say. Surely we must look for a concept of capital, a measure of capital, which is proof against that kind of disturbance.

There seems then to be no alternative but to build up from the business account, instead of the personal account. Assets will therefore need to be valued *separately*. For that is what the company does on its own balance-sheet. It is a valuation more like that on the company's balance-sheet which we seem to require.

We have agreed that the individual 'machine' must have a value which is in excess of its scrap value; can we get any further? In the case of a new machine, we could get further. We could say that the machine would not have been purchased unless its value, to the business purchasing it, was (or appeared to be) in excess of its cost; and we could accept, at least as a convention, that even after its purchase, immediately after its purchase, the same should be taken to be true. We could then go on (much as we did in the case of the marketable security, but not quite for the same reason) to say that the cost of the machine be taken to represent its value to the firm that has taken possession of it.

I

But in the case of an old machine, one that was purchased some time ago, that will clearly not do. For it is usually not possible to buy upon the market a machine the condition of which is identical with the condition of the old machine that has to to be valued; so (if we reject the scrap value) the only value that can be set upon it is a subjective valuation. We shall nevertheless want that valuation to be made in a way which is consistent with the valuation *at cost* of the new machine. This seems to imply that we should take the cost of a new machine, such as could now be purchased, and *write it down* by a subjective assessment of the inferiority of the old machine relatively to the new. This seems to be the nearest thing to the valuation we require that we could conceivably get.

It may appear that in making this prescription, we are not far away from business practice: the valuation of assets at cost, written down for depreciation and obsolescence. There is nevertheless an important difference. The balance-sheet of the company, as explained, is designed to show shareholders what has been done with their money. If it is this which has to be shown, the original cost of the actual assets held is the magnitude that is relevant. As it appears to the accountant, he has two firm values, the original cost and the final value (zero or scrap); he estimates the intermediate values by interpolating between them. But for a correct estimate of the value of a capital stock at a point of time, as it is 'now', original cost must be irrelevant; *bygones are bygones*. It is not what was paid for the machine in the past, it is what would be paid for a comparable new machine in the present, which is the thing that has to be written down.

In what way comparable? What precisely should be meant by the new machine, with which the existing old machine is being compared? It will hardly do to suppose that it must always be a machine of identical technical description. Machines of that description might well have gone out of production, so that if one such machine had to be produced, to special order, the cost would be enormous. It is clearly more appropriate to suppose that (for some reason) the existing old machine has to be replaced; and to take as the cost of the 'new machine' the current cost of whatever machine could now be produced with fair efficiency and which would do (more or less) the same job. This is not a very satisfactory prescription; nevertheless it is in this direction that we seem to be driven to look.

For the carrying-through of a valuation on this basis, there are two ways in which the conventional valuation at original cost (or written-down original cost) would seem to require to be corrected.

In the first place, even if we were to think of the old machine being replaced by a new machine of identical specification (assuming such machines were still in production), the cost of such a machine, now, might well be different from what it was at the date when the old machine was originally acquired. In the inflationary conditions which in so many countries we have experienced for so long, the efflux of time would probably have led to a rise in that cost. The original cost would therefore need to be written up for the rise in price.

In the second place, we should have to allow for technical progress. As a result of technical improvement, the cost of a machine which would do the same job, or an equivalent job, might well be lowered. So the machine, if it were to be replaced, would not be replaced by a machine of identical specification, but by a cheaper or more efficient machine. So the original cost would have to be written down for technical improvement.

Now it may well be said that this second correction, unlike the first, is quite normally made in business practice. For is it anything else but an allowance for obsolescence? It does not arise in the same way as the accountant's obsolescence allowance, which is made by estimating the date at which the existing asset will need to be replaced, not on account of physical deterioration but on account of technical improvement; yet it springs from the same cause, and may (more or less) correspond. It should however be noticed that an obsolescence allowance, calculated on the business principle, covers loss of value due to *probable* future technical progress, as well as that which has actually occurred in the past (between the date when the machine was originally acquired and *now*); but it is only the latter which is relevant to valuation on the principle we are discussing. (Thus there might have been no actual improvement between the acquisition date and the present; the accountant would nevertheless feel himself justified in making an allowance for obsolescence.) But this is a small point, and should probably not prevent us from regarding the loss of value, registered as obsolescence, as being more or less adequate to deal with the second correction. The first adjustment (for change in prices with unchanged specification) is on the other hand not generally made, save sometimes in extremely inflationary conditions; yet for a logical valuation of real capital, it would seem to be at least equally necessary.

To correct the balance-sheets of companies, one by one, so as to make them conform to the principle I have been outlining, would very obviously be a formidable undertaking. One could not expect that it should be done by the accountants, the purpose of whose calculations (as we have seen) is quite different. All that can be done on these lines, in practice, is to follow the principle described, so far as we can, when making statistical estimates.

It is fairly widely held that by indirect methods, by proceeding from information about values of gross investment in past years, applying to them estimates of the 'life' of capital goods and adjusting them by index-numbers of the prices of such goods, a figure which will represent the capital stock of the economy, valued at current prices, can in fact be arrived at. I shall regard it as no part of my task to examine those methods. I shall treat the statistical estimate as an estimate of the figure which would be reached if the balance-sheets of individual firms were corrected on the principle I have been outlining. If we had this 'true' figure, what could it do for us?

What we should like it to do for us would be to provide a measure of the *quantity* of non-human resources inherited from the past and available in the present to be worked on by labour to produce current output. And it may well be said: is not that what it purports to give? The prices used for valuation would all of them be current prices, so it could quite properly be deflated by an index-number of the prices of capital goods, to give a measure of real capital. It would be the new capital goods which would be the part of the aggregate which was most firmly valued, others being valued by reference to them; so it would be an index of the prices of new capital goods which we should want to apply. Such indexes often exist; what is the trouble?

The deflation, I fear, is the snag. Though we have index-numbers of the prices of capital goods, it is uncertain whether they are suitable for the delicate task which is here at issue.

Any index-number comparison, whether it takes the form of a quantity- or of a price-index, involves a comparison between the character of the articles that are valued at one date and at another. In the familiar case of consumption-good price-indexes, we are content to treat the character of the goods in the 'basket' as being the same at the various dates that are under comparison; a cauliflower is a cauliflower, an electric light bulb is an electric light bulb, in 1955 as in 1965. Not only do we treat them as physically identical; it is an acceptable assumption that they satisfy the same want. But when we come to the 'basket' of capital goods, in an economy where technical progress is rapid, these comfortable assumptions fail.

The problem, it will be remembered, is one that we have met before. We met it over the valuation of the capital stock of a single date, considered in isolation. In the valuation of the 1955 stock, considered in isolation, there was already the problem of valuing the old capital goods, produced before 1955, in terms of the new goods produced in that year. Our solution, the solution to which we seemed to be compelled, was to reckon the value of an old machine as the current cost of a new machine, which would do the same job (or which, more generally, would produce an 'equivalent' product), written down for 'inferiority'. That is to say, we translated old machines into new machines by reference to their comparative efficiency or productivity. There did not seem to be any other alternative.

When we come to the 1955-65 comparison, we are faced again with the same issue. We have, by the method indicated, reduced 'old' 1955 machines to new 1955 machines, and 'old' 1965 machines to new 1965 machines; but we have still, if we are to make a *real* comparison, to reduce new 1965 to new 1955 (or vice versa). Logic evidently dictates that we should adopt the same solution for this other problem.

I much doubt if this has ever been done by the statistician; I doubt if he has the wherewithal to do it. But I have also no doubt that it is what the theorist, if he sticks to his theoretical principles, should require to be done. The consequences of adopting the theoretical solution are nevertheless striking. We cannot clear our minds on the issues involved unless we are aware of them.

Capital, if measured in this theoretical manner, is measured, ultimately, by its productivity or efficiency. A capital stock which is capable of producing more (as measured again, of course, by an index-number comparison) is *by definition* a larger capital stock. Thus it is not true, as the older economists used to say, that 'capital is the result of saving'; it is true that it is increased by saving, but it is also increased as a result of technical improvement. It is only in a technologically stationary economy that Mill's Fundamental Proposition would be correct.

The effect of technical improvement, as we have seen, is to lower the value of old capital relatively to new capital. (There are exceptions to this rule, in the case of natural resources, if we regard them as capital; generally speaking, however, this holds.) That is the effect on the existing old capital; it nevertheless follows, on the same principle, that the effect of the same improvement is to raise the value of new capital, of new investment, relatively to the new investment of former years.

So if one takes the former year as base (looking at 1965 in terms of the values of 1955) the new investment of 1965 is put up as a result of technical improvement; while the old capital of 1965 is down in terms of 1965 new capital, but may be either up or down in 1955 terms. The accountant's obsolescence procedure, as we have seen, would always tend to put it down. But it is pretty clear that from the standpoint of the measure we are now discussing, the accountant is too pessimistic. Since he does not allow (and from his own point of view is quite right not to allow) for the increase in the productivity of new investment that comes from technical progress, he overestimates the *gross* investment that is needed to maintain capital intact. Capital (*real* capital) increases faster than it appears to do on conventional accounting methods, by the reinvestment of depreciation (and obsolescence) allowances on continually more favourable terms.

This is not at all the conventional way of looking at the matter; but it can hardly be denied that it makes sense. Not only is it logically consistent; it gives an intelligible account of what we see happening. Why is it that a simple improvement in technical knowledge, even if the knowledge is quickly diffused among those who can take advantage of it, does not (indeed cannot) exercise its full effect upon production all at once? The obvious answer is that the stock of capital (real capital) has to be transformed; it has to be put into such a form that it is adapted to the technical possibilities which have been opened up. The means to the transformation consist of the new capital goods produced, the *gross* investment that is being performed. If gross investment is low, the transformation will be slow and gradual; but if it is high (if it would have been high even without the technical improvement) the improvement itself can be more rapidly absorbed.

That is what would be shown, quite clearly shown, if capital were measured in the theoretical manner I have indicated; but, as I have said, I do not believe that this is a measurement which, save perhaps in some unusual circumstances, is statistically practicable. Even if the difficulty of reducing old capital to new capital at a given date could be overcome (and I have made it clear that I am under no illusion about the difficulty of that reduction), it is nearly inevitable that the only available price-index of capital goods will be biased in the direction of physical comparability. It is bound to have a tendency to measure capital in terms of physical content (sometimes even just 'so much steel'), instead of measuring it in terms of its productive efficiency, as for the theoretical measure would be required. Improvements in the manufacture of capital goods (as in 'steel' production) would be registered; improvements in the utilization of capital goods would slip through.

I cannot myself perceive that there is any economic sense in such a physical measure. It is futile to erect great edifices of theory, and of econometrics, upon it. The estimation of production functions, involving a distinction between accumulation of capital (in some such sense as this) and technical progress (*residual* technical progress), seems therefore to me to be a vain endeavour.

There is one alternative that remains. Why not abandon the capital-price index, and deflate by the consumption price-index, where we are surely on safer ground? In terms of the running account, the income and expenditure account, we are accustomed to thinking of this as a useful procedure. Why not proceed, throughout, in those terms?

GNP, deflated by the consumption price-index, measures the consumption equivalent of gross output; gross investment, similarly deflated, is reduced to its consumption good equivalent. That is to say, it is valued in terms of consumption goods given up to get it; and though it is not to be supposed that consumption could be increased (from available resources) to the full amount of this equivalent, even if gross investment were cut down to zero, the rate at which one can be substituted for the other at the margin (which is what is purported to be shown) has doubtless at least some economic significance. Thus far we could go; and in terms of the running account of the single year (the national income and expenditure account), it often appears to be useful to go so far. It is, after all, what is continually done, if only by implication, for budgetary and such-like purposes.

If the investment that is performed in any year can be represented in this way in terms of the consumption of that year, and if the consumptions of successive years can be reduced into terms of one another by the application of a consumption price-index, cannot the capital existing at any time (at least so far as it consists of reproducible commodities) be represented in the same manner—just by adding up? Is not this a more promising way of getting a figure for *real capital*? True that it is net investment, not gross investment, which would have to be added in this manner; so we are brought back to the question of estimating 'capital consumption'. But that is a problem which, as we have seen, comes up, in some form or another, whatever approach we adopt. It is not obvious that it is more intractable on this approach than on others.

All the same, though not for that reason, I believe that this is an approach which should be rejected. The efforts and sacrifices which were necessary, in the past, in order that some part of the equipment, which the present has inherited from the past, should be constructed, were no doubt of the greatest importance *then*; but what is their present importance? What light do they throw upon the nature of our inheritance, as it now is? The *present* stock is the matter with which a useful measure ought to be concerned. Once again, *bygones are bygones*.

That is in fact the note on which I wish to conclude. Capital is wealth at a moment of time; but time is a flow; we do it violence when we seek to stop it for a 'moment'. Capital has value because it can be used in the future; but who knows, now, what use the future will make of it? Capital goods are in existence because it was thought, in the past, that what was then the future would be able to make use of them; they are inherited from the past, they exist because of the past. They are links between past and future; however we look at them, these links, with past and future, are essential to them. Yet in making a valuation of them, which is itself to appear in a time-series, we must give them *one* time-reference; we must bring them back to the present.

We reject the market value, as determined on the stock exchange, because it is too future-oriented; it is too much affected by hopes and fears for the future, by guesses which are not fact but at the best probability. We reject the 'original cost' of the balance-sheets of businesses, because it is too much oriented towards the past. We are baffled in our pursuit of a consistent 'national balance-sheet', because it is impossible to reduce these forward-looking, and backward-looking, valuations into even a formal consistency. We may doctor them to make them appear consistent, but how much do we gain when we do so? (I must say that if we have to doctor, I would prefer to doctor the business balance-sheets than the personal balance-sheets. Businesses, after all, must make forward valuations, though they do not publish them; they make their decisions with respect to the future, not with respect to the past. It is only for the curious purposes of a capital gains tax that the shareholder need concern himself with backward-looking valuations.)

The statistician's job, if he attempts a measure of capital, is to bring these practical valuations, that are given in his data, too forward-looking on one side, and too backward-looking on the other, into something which should belong, as much as it can belong, to the moment of time, *now*. It cannot be expected that he will ever be very successful.

RESUME

Le capital national peut être évalué en faisant usage: (1) des prix des actions sur le marché, (2) des coûts des biens capitaux qui figurent dans les bilans des entreprises. Pour ce qui regarde la distribution des richesses le premier moyen est de rigueur; pour mesurer le capital réel, il faut avoir recours à quelque chose qui dérive du deuxième. Mais il n'est pas facile de trouver un moyen de rapporter les valeurs des capitaux anciens à celles des capitaux neufs, et de rapporter de la même façon les valeurs d'une année à celles d'une autre. On doute que cette difficulté puisse être tout à fait résolue.

REFERENCE

HICKS, J. R. (1952). *Social Framework*. 2nd ed. Oxford.

DISCUSSION

SIR ROY ALLEN. I wish to underline what the author has said on the link between the definition and estimation of macro-economic aggregates. Estimates to fill gaps in knowledge of records are one matter; estimates of things which are not valued or recorded are another matter, i.e. the problem of imputation. National income aggregates illustrate this. They can be defined to avoid imputation but useful definitions include at least some imputation, e.g. rents of owner-occupied houses and, more important, values of changes in stocks. Turning to aggregate capital, we find that the problem is almost entirely one of imputation; these are seldom actual transactions and not often agreed valuations of capital stock. The computation of a capital aggregate is not worthwhile unless a clear and appropriate definition of capital is provided. There are at least three possibly useful definitions. One is the aggregate money value (and distribution) of property and this has uses, at least in occasional computations, to supplement the distribution of incomes. Another is the capital accumulation of industry in real terms and this has not yet established its validity for use in time-series estimation of production functions. A third is also a real concept of capital but as a consumption goods equivalent and this is the measure which the author regards as most useful, e.g. for capital-output ratios.

R. ROY. Je présenterais seulement quelques remarques générales sur l'importante communication de Sir John Hicks.

1. Le calcul d'un Indice, représentatif de la valeur d'un Capital Productif, dépend de l'objectif que l'on s'assigne.

Si l'on entend par exemple résoudre un problème relatif à la croissance optimale de l'économie, on peut légitimement se référer à la valeur de la consommation au cours de la période considérée.

2. Dans cette optique, les matériaux puisés dans la Comptabilité Nationale peuvent utilement servir pour le calcul numérique de l'Indice représentatif du Capital de Production. Nous n'insistons pas sur les difficultés que soulève à cet égard la variation possible du pouvoir d'achat de la monnaie.

3. La liaison entre consommation et valeur du capital nécessaire à sa production soulève une difficulté particulière: la production possible, pour un capital déterminé, se trouve-t-elle représentée par la capacité de production qui constitue un potentiel maximum ou par une valeur moyenne, définie au cours d'une certaine période et, tenant compte des mouvements de la conjoncture?

Nous signalons, sans y insister, les difficultés auxquelles donnent lieu le calcul de l'amortissement d'un capital et des phénomènes d'obsolescence qui s'y rattachent.

C. CLARK. In making any international or long-period comparisons of capital stock we must recognize that some two-thirds of capital stock consists of structures whose price moves differently from the average, because construction is a labour intensive industry, whose productivity rises only to about the square root of the rate of rise of general productivity. The purchasing power of money over structures in the poorest countries may be five times as high as in the wealthiest, and the real capital stock of the poor countries is consequently underestimated. There is clear evidence from all countries of increasing efficiency in the use of capital, particularly in the use of structures, provoked by their rising relative cost. Floor space requirements per unit of manufacturing product in U.S.A. have been halved over the last four decades. There are signs however that this process is now slowing down. Short-period fluctuations in gross profits provide a good measure of pressure of current demand on net capital stock and so provide a good check on estimates of the latter. Professor Jorgenson is undoubtedly right in pointing out that when we construct production functions what we really need is not the *stock* but the *flow of service* from capital. A dollar's worth of equipment which is quickly obsolescing is giving about twice the flow of service of a dollar's worth of structures.

The Concept of Income in Relation to Taxation and to Business Management

John Hicks

The subject of this paper is one with which I have been concerned on two quite separate occasions during my life as an economist. On each occasion I have had to think about it, but I have had to think about it in quite different ways. One was when I was writing the chapter on Income in my *Value and Capital* (1939), a chapter which has since been regarded, at least in some circles, as a rather standard presentation of the theory of the matter; the other was when I was a member of the British Royal Commission on the Taxation of Profits and Income (1950–53). What I mainly learned on the latter occasion was the remarkable difference between the meaning (or meanings) of income, as the notion appears to the economic theorist, and the meaning which is given to it, which indeed, as I shall be explaining, has to be given to it, when income is considered as an object of taxation. Nevertheless, though the meanings are different, they are related. Thus when the fiscal concept gets into difficulties, as it notoriously does in times of inflation, these difficulties raise issues which take one back to the theoretical concept. For it something much nearer to the theoretical concept which, for his own purposes, the business man requires.

It is revealing to look at the issue historically. I think it is generally agreed that the ancestor of all modern income taxes is that which was imposed, in England in 1799, for the financing of the war against the French Revolution. It has often been said that this was a combination of five separate taxes, the "schedules" which have been a formal feature of the British income tax ever since. The point was that incomes of different sorts had to be reached in different ways. It was easy to tax the interest on government debt, for the facts of that were directly known to the government; taxes on income from land were familiar, and could continue to be assessed in the old way; contractual salaries were fairly easily assessable, by reference to the contracts by which they were paid. Where a difficulty arose was in Schedule D, business income. What was the income of a business man, say the owner of a cotton mill? (I take what would have been a leading case in England at the time in question.) All that could be done, to begin with, would be to get him to state it himself ("self-assessment"), with just a hope that he could be shown up, from some sort of common knowledge, if what he declared seemed too outrageous.

73

For the fact was, at that stage, that one could not even be sure that the business man himself would know what his income was. When he was bidden to declare his income, he might well reply: what do you mean? And what answer could be given to him? Could anything be done but to refer to the old principle — already, at that date, an old principle — of the desirability of "living within one's income"; if one spent more than one's income, one's prospect for the future would be damaged. So, on this line of thought, the income derived from a business would be the *maximum* that could be *safely* taken out of the business, for personal expenditure or for the payment of taxes, without damaging the prospects of the business. But that, it is clear, would be a matter of judgement; just how much is *safe?* The tax-collector would be in no position to form much of a judgement on the matter; so he would have to rely, very largely, upon the taxpayer's judgement, who could not be prevented making a judgement very much in his own favour.

It was therefore inevitable, so long as this situation persisted, that income tax would bear less heavily upon the profits of business, than upon other incomes which could be more objectively assessed. It just could not reach profits in the way it could reach other incomes. An income tax with this deficiency could not be regarded as a fair way of taxing; and it is clear from the evidence that it was not regarded as fair. Later on, in England at least, but I suspect that the same story has been repeated in other countries, there has been a change.

I believe that the main cause of the change has been the development of the limited liability company, or corporation. A shareholder in such a company is nominally a part-owner of the company; but his situation is more realistically described if he is thought of as one who has lent money to the company, not on the promise of a fixed payment of interest on the loan, but on the promise of a share of profits. When the usual method of raising capital is by borrowing at fixed interest, no more need be required by the lender than that he should be assured of the general "soundness" of the borrower; it is not necessary for him that the profit which is to be earned by the borrower should be exactly assessable. But if what is promised is a share in profits, the prospective shareholder has the right to enquire: what profits? So the growth of the company form of organisation required the development of company law, including regulations for the ascertainment of profits, in what could at least be made to appear to be an objective manner. This put the tax-collector in a much better position, for he now had an ally. There was the shareholder, with another interest in the ascertainment of profits, who came to his assistance.

At this point the accountant enters. He has now to be given a regular place in the ascertainment of profit — a duty to provide a fair account of the performance of the business, for the benefit, in the first place, of the shareholders; but the work which has been done for the benefit of shareholders can also be used for the assessment of tax. He will naturlly desire to be as objective as possible, in order to minimise disputes between the (several) parties that are now concerned. But this means that he cannot measure profit, or income, in the former sense, of the maximum that can be *safely* taken out of the business, for on that there can be endless argument. He must seek to measure profit in some other way. It was fortunate, for him, that there was another way to which he was already accustomed.

One does not need to have much detailed knowledge of the history of accounting to see that its origins must go back to a time when industry was rudimentary; so that the main kind of activity for which accounts were required was merchandising. There are several ways in which the terminology of accounting bears traces of its mercantile origin. "It is characteristic[1] of the business of the merchant that it is divisible into separate units. Every bale of cotton or pound of cheese which ever forms part of his stock is acquired at a particular date and sold at a particular date: purchase, retention and sale constitute a separable transaction. (Complete separation is of course not attained in practice, since there are overheads which have to be allocated; but it is so nearly attained that it sets the pattern.) So the accounts of the merchant may be regarded as a bundle of separate accounts. Purchases and sales are indeed going on continuously, so that if they are set out in a time-sequence, the separate accounts will overlap; it is only if an account were prepared for the whole history of the business, from first setting-up to final closing-down, that a record of purchases and sales would tell the whole story. In any annual period there must be transactions which have started before the beginning of the year but are completed within the year; there must similarly be transactions which are begun within the year but are not completed at the end. But these in the case of a mercantile business cause the minimum of trouble. They can be dealt with by the accepted rule of never taking a profit on any transaction before it is completed. The initial stock of the year will then be brought in *at cost*; and the final stock will be valued at cost in the same manner.

Now what was to be done, on these principles, with plant and machinery? The use of land, being regarded as a permanency, could be brought in as a regular charge; but the plant and machinery is not expected to last indefinitely, though its use is spread over a time which is longer than the accounting period. It is important to observe that it is the extension of the use to a duration which is longer than the accounting period which creates the difficulty. There would be no difficulty, here as in the mercantile case, if the account were drawn up for the whole life of the business, from first setting-up to final closing-down. It is for the annual account that there is the problem. The cost of the machine has to be set against a series of sales, the sales of the outputs to which it contributes,

---

1. I am borrowing this passage from a paper of my own on *Capital Controversies: ancient and modern*, where I had to make use of this analysis of accounting procedure for another purpose. (The paper is re-printed in my *Economic Perspectives*, 1977; the passage in question is on pages 157–8.)

but some of these sales are sales of the present year, some are later, and some, maybe, earlier. There is thus a problem of imputation; how much is to be reckoned into the costs of this year, and how much into the costs of other years? It is just the same problem as the allocation of overheads, and to that, as is well known, there is no firm *economic* solution.

Neither has the accountant found a solution — only a name and a set of, essentially arbitrary, rules. The "depreciation quotas" must add to unity, but that is all that is known, at all firmly, about them. The form of the account is preserved, but only by bringing in, as the capital which is supposed to be invested in the machine at the beginning of the year, that part which has not been absorbed (by being allocated as a cost to the outputs of preceding years) and by reckoning as the capital invested at the end of the year that part which has not been absorbed in those years nor in this year — so that it is left to be carried forward to the future. This is in fact what accountants did, probably what they had to do, as soon as they were confronted with the problem. It was what they still do, even to this day."

That the distribution of the accountant's depreciation quotas is essentially arbitrary is fairly well accepted nowadays; thus it has been discovered that it is open to the taxing authority to impose a different rule for the assessment of profits that are liable to tax. (Accelerated depreciation, and so on.) There is thus no reason why there should be any simple rule which would cause the profits that are calculated by its use to have any correspondence with the income that would be assessed by the criterion with which we began — the maximum that can be *safely* taken out of the business. If we call the depreciation which would fall to be deducted from the receipts of the year, on this criterion, the *true* depreciation, the accountant's depreciation is at best a proxy for it. But, as we have seen, the true depreciation is a matter of judgement, so it provides no firm figure. Yet the question still arises — if the true depreciation could be assessed objectively, what relation to the accountant's depreciation would it have? This, we shall find, is by no means a nonsense question. There are some quite interesting things to be said about it.

It will be useful, for this purpose, to write out the accounts of the business in formal terms. Let R be the current receipts of the business during the year, E current expenses, C capital expenses. All these, we may take it, are quite objective. (There is indeed a question of what expenses are to be reckoned as capital expenses, but I shall come to that later.) In contrast with these "firm" items, stand the initial capital, the final capital, and the depreciation, all of which are determined by the accountant in accordance with his accounting conventions. So I shall mark them with stars. K_0^* will thus be the initial capital, as it "stands in the books", K_1^* the final capital and D^* the depreciation.

Then what the accountant presents are
(1) a current account, which could be written

Current profit $= R - E - D^*$

(2) a capital account, which would be written

$$K_0^* + C - D^* = K_1^*$$

From the capital account, given K_0^* and K_1^*, D^* follows; and then, from the current account, current profit is determined.

Now how is one to construct a "true" set of accounts, with which this conventional account could be compared? In general, as I have insisted, it cannot be done, since what would have to replace K_0^* and K_1^* are no more than subjective estimates. There is however an exceptional case, an improbable case, but nevertheless a possible case, in which it could be done. This is the case of the business which had changed hands at the beginning of the year, having been purchased for a price which we will call K_0 (unstarred, since there is nothing conventional about it); and which is then sold again at the end of the year, for a price which we may similarly call K_1. I shall make a good deal of use of this "double take-over" model, which I find to be a convenient tool to bring out principles.

There can, I think, be little doubt that an accountant, who was asked to do the accounts of a business with this peculiar history, would refuse to do them in terms of K_0 and K_1; he would insist in doing them in terms of K_0^* and K_1^*, the values which "stand in the books". The economist, however, would find K_0 and K_1 much more interesting. He would want to start from K_0 and K_1 (doctoring them, maybe, for changes in the value of money, but that again I shall come to later); it would be these *market values* which he would want to take as representing the initial and final capital.

With a double take-over, K_0 and K_1 would be values that would be available to the accountant; what could he make of them? Though he refused to use them for the construction of a capital account *of the business*, he could not refuse to use them for a capital account *of the proprietors*, the temporary proprietors, the proprietors during the year. But in that account, if K_1 stood on one side, and $K_0 + C - D^*$ on the other, it is unlikely that there would be a balance. There would be a difference, which he would reckon as a capital profit, a capital gain (or loss).

It will be noticed that the sum of current profit and capital profit is

$$(R - E - D^*) + K_1 - (K_0 + C - D^*) = R - E - C + K_1 - K_0$$

from which all starred items have disappeared. The distinction between C and E (current expenses and capital expenses) has also become immaterial.

If we denote the "economist's" depreciation by D (suppressing the star), the depreciation, that is, which would be calculated by using K_0 and K_1 as initial and final capital, his capital account would have K_1 on one side and $K_0 + C - D$ on the other, which would balance, since D had been chosen to make it balance. So he would show no capital gain; the whole of what the accountant reckoned as current profit + capital gain would appear as (undifferentiated) profit.

Now it is surely a question whether it is not a weakness of the procedure which has just been attributed to the economist, that it can make no distinction between current profit and capital profit. It is not just that such a distinction is commonly required by legislation; if it had not been the usual practice to pay attention to it, legislators and lawyers would not have paid attention to it — so we must look for a further reason. The events of the year, even from the point of view of the temporary proprietors, have been of two kinds. There has been (1) the regular running of the business and (2) the double take-over. It is not mere curiosity to wish to know how much of the total profit is to be attributed to each.

For one is brought back to the underlying question: why should anyone (apart from tax law and company law) want to have a *current* account of a business? At the time when it is drawn up, it relates to the past; nothing can now be done about the past: "bygones are bygones". The only purpose of having an account of the past is the guidance one hopes to get from it about future performance. The future is unknown, but action about it has to be taken, action that needs to be based on some evidence. The purpose of the current account is clarified when we look on it as a way of presenting some of that evidence.

This is the reason why the separation of current profit from capital profit is important. It is the current profit which has been earned in the past which is relevant to the future; the capital profit is *necessarily* irrelevant. That can readily be seen in the case of our two take-overs. Those who purchased the business in the second take-over will have wanted to look at the current profit, as part of the information on which to base their decision on how much to offer. A capital profit, accruing to the temporary proprietors (those of the year) would depend on what they in their turn had paid for the business in the past; but that, from the point of view of those to whom they sell, does not signify. What concerns them is the value of the business now, not its value at a date some time ago.

The current profit, when it is looked at from this point of view, should be an index of the performance of the business during the year to which it refers; elements which refer to its activity in previous years should in principle be excluded. The accountant's proxies, as is now notorious, do not succeed in excluding them. The D^* which he uses for depreciation is not an estimate, based in some way or other on the actual performance of the business during the year in question; it is derived, by conventional rules, from figures that are given to him, firm figures indeed, but figures of previous expenditure, which do not belong to the year for which the account is being made. In stationary conditions, when one year is much like another, that may not much matter; but in non-stationary conditions it is bound to be misleading.

Is there any way of cutting out this reference to the past, and still identifying a *current* profit? Keynes sought to do it by substituting "user cost" for depreciation; but it is hard to see that user cost can ever be more than a subjective estimate. It does not make contact with the accountant's problems and therefore with tax problems, so it is not much help to us here.

Much more help can be got from the ingenious analysis of Erik Lindahl, published[2] as long ago as 1933. It may be translated into terms of our double take-over model as follows.

We start from the capital account (of the "proprietors")

$$\text{Capital profit} = K_1 - (K_0 + C - D^{**})$$

in which I write D^{**} for D^*, as a sign that the accountant's way of measuring depreciation is no longer to be taken for granted. The first question to be decided, on Lindahl's approach, is to find a way of defining D^{**}, or rather $C - D^{**}$ (the *net investment* of the economist) so as to make the new capital account as meaningful as possible.

Let us re-arrange this capital account, writing it

$$K_1 - K_0 = (C - D^{**}) + \text{capital profit}$$

This shows at once that the problem is one of dividing $K_1 - K_0$, deciding how much is to be allocated to $(C - D^{**})$ and how much to capital profit.

K_0 has been defined as the price that was paid for the business at the beginning of the year. It may thus be regarded, in the manner that has become familiar, as the capitalised value of expected future returns, in what we are regarding as the current year *and afterwards*. K_1 is also a capitalised value, but is relates to a different future. The future, for K_0, includes the activity of the year, but from the future for K_1 that is excluded. This is one difference, but there is another — a difference of information. In the course of the year new information about the prospects of the business after the year is likely to have become available.

The account is drawn up, as is obviously necessary, after the end of the year; it is reasonable to take it that it is the prospect as it appears at the end of the year which is the last to be relevant. At that date the information implied in K_1 is already available. So K_1 is in a double sense a firm figure. It is not just given (by our double take-over assumption); it is also based upon the information available at the exact time at which (or for which) the computation is being made. But when this aspect is taken into consideration, it becomes apparent that K_0 is not a firm figure in this second sense. At the time when K_0 was established, some of the information on which K_1 is based was not yet available.

Consequently, if there is to be parity of information (and the account is to refer to information as it is at the end of the year) K_0 needs to be doctored. We have to consider what K_0 *would have been* if the information available at the end of the year had been available to it. This information includes (1) the actual

---

2. "The Concept of Income" included in *Economic Essays in Honor of Gustav Cassel* (1933).

performance of the business during the year, and (2) K_1 itself. At the end of the year these are known, so K_0^{**}, as I shall call the new substitute, is what K_0 would have been if the actuals of the year (R, E, and C) had all been known, and K_1 had also been known. Though at time 0 these are all in the future, they must all be supposed to be known, with no uncertainty about them. Thus K_0^{**} is just what would be paid, at time 0, for the right to receive R during the year and K_1 at the end of the year, on condition that E and C, during the year, are paid out. The contract in these terms is quite precise, with no uncertainty about it.

The introduction of K_0^{**} enables the capital account, as written above, to be divided into two parts

$$K_1 - K_0 = (K_1 - K_0^{**}) + (K_0^{**} - K_0)$$

Can we identify these two parts with the two parts of our former division, making $C - D^{**} = K_1 - K_0^{**}$ and $K_0^{**} - K_0$ the capital profit?

To interpret the capital profit in this way, making it simply a measure of the gain (or of course loss) due to change of information between time 0 and time 1, seems quite agreeable. One does want to think of it as a revaluation, due to information, and that is what is shown. But the consequence, on the side of current profit, is, at least at first sight, surprising.

For the "contract", as explained, is precise; it is so precise that it is possible to calculate K_0^{**}. We have only to introduce a rate of interest (r), which must be supposed to be established outside the business, and perform the discounting. Since the contract is certain, a gilt-edged rate of interest, with no risk-premium in it, will do. All that then remain to be specified are the dates, within the year, at which R accrues, and at which E and C are paid out. Nevertheless, because of the given rate of interest, these can all be accumulated, or discounted, to a single date, and we may suppose that that has been done. So let them be represented by their values at the *end* of the year. Then

$$R - E - C + K_1$$

is the value of the contract, at the end of the year; and this is to equal $(1 + r)K_0^{**}$.

Now see what happens when we put $C - D^{**} = K_1 - K_0^{**}$. Current profit $= R - E - D^{**} = R - E - C + K_1 - K_0^{**}$
$= (1 + r) K_0^{**} - K_0^{**} = rK_0^{**}$

So current profit is just the interest on the calculated value of the initial capital — interest that has accrued during the year.

Lindahl himself was well aware that this was the consequence of his proceeding; but what are we to make of it? We have been looking for a definition of current profit which, so far as possible, should register the performance of the business within the year, excluding what has happened before and what is to come after. But

$$rK_0^{**} = (\frac{r}{1 + r}) (R - E - C + K_1)$$

so that K_1, which is the capitalised value of expected net receipts *after* the year, would appear to have a large part, even, in many cases, the dominant part, in determining the *current* profit. So the Lindahl definition, when it is looked at in this manner, seems still to be infected by the future, having little to do with the performance of the current period.

There is however another way of looking at it. We are still defining current profit as $R - E - D^{**}$; and it follows from the definition of D^{**} that if it were to be the case that $C = D^{**}$, K_0^{**} would equal K_1. Thus we can alternatively define D^{**} as the capital expenditure which would have to be incurred if capital were to be maintained intact, in the sense that the prospect of the business was as good at the end of the year as it was at the beginning, *as we can now see with hindsight that it was at the beginning*. That does look like a formulation which belongs to the current period, as rK_0^{**}, which we have seen to be equivalent, apparently does not.

It will be noticed that in this case, if $C = D^{**}$, and so $K_0^{**} = K_1$, current profit $= rK_0^{**} = rK_1$. So if $C = D^{**}$, so that capital is (just) "maintained intact", the business can look forward, on the basis of information as at the end of the year, to earning in future years a current profit which is equal to that which it has earned in the current year. Indeed, if a sum of money equal to K_0^{**} had been invested in a security with no terminal date, yielding a fixed rate of interest r, rK_0^{**} would be the income that would be derived from that security *in each year*. So rK_0^{**} is the income, constant over time, which is equivalent, in terms of capital value, to that which could have been expected from the business, if what happened during the year had been foreseen at the commencement, and if the knowledge of its further prospects, which is present at the end of the year, had been available at the beginning. This is effectively what Friedman would call the *permanent income* derived from the business. It is to a concept of permanent income that the Lindahl analysis seems finally to come.

If a permanent income is to be derived from a business, it must surely be a continuing business; this is confirmed by what may well appear at first sight as a paradoxical consequence of the Lindahl definition. For suppose we apply it to the case of what is not a continuing business, a business that is set up at (or after) the beginning of the year and closed down at (or before) its ending. Receipts and expenses may be accumulated to the end of the year, as before; so the total net profit, in the notation we have been using, is $R - E$. (Since there is no sense in capital expenditure for the benefit of future production, when there is to be no future production, we may take it that $C = 0$.) How much is current profit, and how much is capital profit? The accountant would surely say that all of $R - E$ is current profit; there is no carry-over from the past, so his D^* must be zero. Keynes, I think, would take the same view; since there is no carry-over to the future, there can be no user cost. But that would not be Lindahl's view. He would say that since the venture is terminated, $K_1 = 0$;

and, as we have seen, $C = 0$. But K_0^{**} would not be zero; with K_1 and C both zero, K_0^{**} would be equal to the discounted value of $R - E$. Current profit would be interest on this; the remainder (with moderate interest much the greater part) would be capital profit.

This follows, inevitably, from the identification of current profit with permanent income. The only way in which a permanent income can be got from a non-permanent venture is by investing the proceeds at interest.

To this, one may be sure, the accountant would protest: the income in question is not an income of the business since it arises outside the business. It is, at the most, an income of the proprietors. As we have ourselves seen, there are reasons why it is important, for accounting purposes, to ascertain an income *of the business;* we must accept that by the Lindahl method this has not been done. Yet I do not think that what I have said about the Lindahl method has been in vain.

For it has enabled us to distinguish, more sharply than could be done in any simpler manner, between the two different distinctions which are hidden away in what purports to be a single distinction — that between current profit and capital profit. There is (1) the distinction between profit that is earned within the business and that which is earned, in some sense, outside it; let us call this the distinction between *internal* and *external* profit. There is (2) the distinction between *normal* profit and *exceptional* profit, which is so sharply identified in Lindahl's analysis. These two distinctions are entirely different. It may happen that an external profit is exceptional; it may happen that an exceptional profit is external; but there is no reason of principle why the two distinctions should coincide.

When a machine, only part of the capital cost of which has been written off, is sold by the business that has been using it for a price which is greater than that at which it "stands in the books", the difference is reckoned by the accountant as a capital profit. He has a reason for doing this. He regards the profit as external; the firm is not in business to sell used machines. Gains that have been made in this way are irrelevant to the valuation of the firm as a "going concern". Once we have grasped the distinction between external and exceptional, we can understand that there is sense in the accountant's convention.

The economist however, who is not particularly concerned with whether the profit is earned inside or outside, is, again quite properly, more interested in the normal-exceptional distinction. For the business itself, both are important; but one would think that it is the normal-exceptional distinction which is the more important. For it is this latter which corresponds most nearly to the subjective notion with which we began — of what can safely be taken out of the business for personal expenditure (or distribution of dividends) and for payment of taxes. A company is usually restrained by law from paying dividends "out of capital"; so it is common practice to accumulate a reserve fund, or dividend equalisation fund, from which it can pay a "normal" dividend even in years when profits have been *exceptionally* low. The fact that this is so commonly done is evidence that the conception of a normal profit, in the temporal sense that here is relevant, does play a part in business policy.

Now what is the bearing of all this on taxation? I will be bold to say that I see no case for the treatment of external gains, for tax purposes, any differently from internal gains. The purpose for which they are so treated, by the accountant, seems to be irrelevant to the question of tax. This holds, not only for the taxation of companies, but also for the taxation of persons. In the British income tax, gains from betting, or from prizes (such as a Nobel prize) are excluded from tax, essentially on the ground that they are external, not being derived from the business from which the taxpayer earns his living. Gains from stock exchange speculation are taxed separately, essentially on the same ground. When I say that the internal-external distinction is irrelevant, I imply that all these should be taxed as income. And so — an even more startling conclusion — should gains from bequests.

Any widening of the scope of income tax, in this direction, must however make more critical the question of exceptional gains. For though the distinctions are different, there can be no doubt that in practice there are many external gains which do cause income (in the inclusive sense including external gains) to become *very* abnormal. If bequests — to take the leading example — were taxed as income, the income of the taxpayer, in the year in which the bequest was received, might well be many times his normal income, even his normal income after he had received the bequest. If income tax were proportional, imposed at the same rate whatever the income, this might not much matter. But no income tax is ever completely proportional; there is always an exemption limit and there usually is progression above it. To tax a large *exceptional* gain, accruing in one year only, having little or no relation to the taxpayer's normal circumstances, on a progressive scale, would surely be monstrous. A widening of the tax base to include external gains, if it were to be at all comprehensive, would therefore have to be accompanied by special treatment of exceptional gains.

That could not, I think, be an easy matter. One can think out schemes for doing it, but they get into great difficulties if they are carried very far. It would be inappropriate, when concluding this paper, to dilate upon them. Instead, I shall make two general points.

First, I am not meaning to suggest that a simple way, which may be thought to be suggested by the formal analysis I have taken from Lindahl, is in fact a solution. If "true" income is interest on capital, as Lindahl appears to say, and if the rate of interest which is to be applied can be assumed to be given objectively, would not a tax on capital value be the best way of taxing income from capital? If it were to be a continuing tax, it would have to be applied at a rate which was less than the assumed rate of interest; thus if the rate of interest were 10%, a capital tax of 3% would be equivalent to an income tax at 30% per annum. Replacement of tax on income from capital by a capital tax is a project which naturally appeals to economists; but I would not wish to be thought to be recommending it myself. It is not just that the cost of repeated revaluations would be enormous; the principal objection is that the capital value of assets, which are not currently being sold, is a matter of judgement, or of opinion. The accountant's conventional way of measuring income, though (as we have seen) it has elements in it that are arbitrary, is largely based upon actual transactions, those which we have been symbolising by R and E. It would not be an improvement to replace this relatively firm assessment by one that in practice must be even more, even much more, arbitrary.

I avoided this difficulty in my model, by assuming that the business, the profit of which was being assessed, did actually change hands, in a take-over. When that occurs, it gives a *true* capital value; but (fortunately) it does not often occur. The mere purchase and sales of shares, which do not carry control with them, does not establish a capital value in the relevant sense.

The other point I want to make concerns inflation. If income from capital could be taxed by taxing capital, the tax, at least in principle, would be inflation-proof; for the assessment would be made, and the tax would be paid, in money of the same value. (I am of course abstracting from lags in the machinery of assessment and collection.) A conventional income-tax, as we know so well nowadays, is not inflation-proof, since the values that "stand in the books" are based on costs that were incurred in earlier years, when money did not have the same value. It would be out of place, at the end of this paper, to say anything much about the proposals, now so widely canvassed, for mending this distortion. I will merely observe that the conventional accountant's method, even when it is applied in conditions of fairly stable prices (of final products) aims to do no more than to identify the surplus, above what is needed to maintain capital intact, when capital is interpreted as money capital; but this implies that if, with economic progress, the prices of "machines" are falling, relatively to the prices of finished products, more will be deducted, on the accountant's method, than is necessary to keep the business in working order as a going concern. Perhaps it is just as well that that is so; it gives an incentive to take advantage of technical improvements! It follows however that if, in times of inflation, all that is desired is to give depreciation allowances which will have the same effect as those which were given in times of constant prices, the answer should be to write up the values carried forward by a *general* price-index. Yet it seems inevitable that when it is proposed to make corrections to the conventional allowances, to adjust to inflation, the question of whether the conventional allowances were appropriate, even in absence of inflation, is bound to be raised. So one comes back to the subjective assessment — how much can be *safely* taken out of the business — from which the accountant's procedure had been thought to be an escape. There are many things which inflation throws into the melting pot, and this is one of them.

Résumé

Le sujet principal traité dans ce rapport est la relation entre le profit courant et le profit en capital (ou gain en capital). L'on démontre que cette distinction peut être réalisée de deux manières différentes et que ces deux interprétations sont généralement confondues. Il y a (1) la distinction entre profit *interne* et *externe*, le premier résultant de l'activité d'exploitation et le second lui étant en quelque sorte extérieur; et (2) la distinction entre profit *normal* et *exceptionnel*, le premier correspondant à ce que Friedman appelle le "revenu permanent" et le second étant constitué d'écarts par rapport à ce premier. Chacune de ces distinctions a des rapports avec la pratique des affaires, et demande à être examinée à la lumière de l'impôt.

L'auteur ne pense pas qu'il y ait lieu de traiter fiscalement les gains externes différemment des gains internes. Et ceci vaut non seulement pour l'imposition des affaires, mais aussi pour celle des personnes. Le seul fait qu'un gain provienne de sources différentes de celle dont le contribuable tire son revenu principal pour vivre, n'est pas une raison suffisante pour traiter ce gain séparément. Cependant, de nombreux gains externes sont exceptionnels; et du fait de la progressivité de l'impôt sur le revenu, les gains exceptionnels appellent un traitement exceptionnel.